Small Particle Ring Accelerators and Paul Traps

Case studies and prospects

Online at: https://doi.org/10.1088/978-0-7503-4891-1

Small Particle Ring Accelerators and Paul Traps

Case studies and prospects

Santiago Bernal

Institute for Research in Electronics and Applied Physics, University of Maryland, College Park, MD, USA

IOP Publishing, Bristol, UK

ISBN 978-0-7503-4891-1 (ebook)
ISBN 978-0-7503-4889-8 (print)
ISBN 978-0-7503-4892-8 (myPrint)
ISBN 978-0-7503-4890-4 (mobi)

DOI 10.1088/978-0-7503-4891-1

Version: 20231001

IOP ebooks

British Library Cataloguing-in-Publication Data: A catalogue record for this book is available from the British Library.

Published by IOP Publishing, wholly owned by The Institute of Physics, London

IOP Publishing, No.2 The Distillery, Glassfields, Avon Street, Bristol, BS2 0GR, UK

US Office: IOP Publishing, Inc., 190 North Independence Mall West, Suite 601, Philadelphia, PA 19106, USA

In memory of my brother Alejandro Bernal (1960–1976)

Contents

Preface

This reference text covers the description, theory and history of room-sized or smaller ring accelerators and Paul traps. The book describes four distinct small ring accelerators and three linear Paul traps, as case studies of scaled experiments for the exploration of advanced concepts and techniques in accelerator and beam physics with potential applications to larger planned or existing machines. The four rings are: 1) UMER, a small electron storage ring at University of Maryland; 2) IOTA, a larger electron/proton ring at Fermilab; 3) SIR, a compact ion ring accelerator, originally at Michigan State University; and 4) EMMA, a model fixed-field alternating gradient machine (a cyclotron/synchrotron-like device) that operated in the UK. The linear Paul traps were employed in the U.S., japan and the UK as RF (radio-frequency) very compact simulators of accelerators. Appendix A summarizes basic scaling laws applicable to ring accelerators, for both synchrotron light sources and colliders. Appendix B contains a brief description of computer resources and examples for modelling accelerators.

Highlighting the work of small groups at universities and national labs, this book brings together material from many publications from conference proceedings and specialized journals for a broad audience of students and accelerator professionals. Although the book is largely self-contained, it refers often to another book by the same author: *A Practical Introduction to Beam Physics and Particle Accelerators, Third Edn.* (IOP Publishing Ltd, 2022). In fact, chapter 9 of the latter is a condensed version of the present book.

Acknowledgements

Special thanks to our colleagues and current and past collaborators at the University of Maryland: Dorothea Brosius, Max Cornacchia, Irv Haber, David Sutter, and Professors Tom Antonsen, Brian Beaudoin, Rami Kishek, Tim Koeth, and Pat O'Shea. We also give recognition to the hard work of several former students in the UMER project: Heidi Baumgartner Komkov, Levon Dovlatyan, Samuel Ehrenstein, William Hartmann (R.I.P.), David Matthew, Sarvenaz Memarzadeh, and Kiersten Ruisard. We thank Laurent Deniau (CERN) for help with the MAD-X code (Appendix B), Levon Dovlatyan (Raytheon Technologies) and Eduard Pozdeyev (Fermilab) for granting us permission to use material from their PhD theses. In addition, thanks to Michael Borland (Argonne Natl Lab) for allowing us to use screenshots of the TAPAs application, Giulio Stancari (Fermilab) for his photo of IOTA and for directing us to important publications, and Fiza Mulla (Freedom High School in Virginia) whose summer project in 2020 (via Zoom) on scaling accelerators was the inspiration for the book.

We also acknowledge the patience and support of our family and the publishers.

Finally yet importantly, we acknowledge the continuing financial support of the Offices of Science and High Energy Physics of the U.S. Department of Energy.

Author biography

Santiago Bernal

Santiago Bernal obtained a BS in physics from the Universidad Nacional de Colombia in Bogotá, Colombia in 1981, an MS in physics from Georgia Tech in 1983, and a PhD in physics from the University of Maryland, College Park in 1999. He worked for his PhD under the direction of the late Prof. Martin Reiser. Before his graduate studies at Maryland, Bernal taught college physics and math in Colombia and Puerto Rico. Bernal joined the UMER group in 2000 as a postdoc, later becoming a research scientist at the Institute for Research in Electronics and Applied Physics (IREAP). He was involved in the design and construction of UMER and has been a leading experimentalist in the project. Bernal helped organize and conduct an experimental course in UMER for the 2008 US Particle Accelerator School. He is the author of *A Practical Introduction to Beam Physics and Particle Accelerators, Third Edn.* (IOP Publishing, 2022). Chapter 9 of the latter formed the basis for this book.

IOP Publishing

Small Particle Ring Accelerators and Paul Traps
Case studies and prospects
Santiago Bernal

Chapter 1

Space charge and other physics at the University of Maryland Electron Ring (UMER)

We discuss in this chapter research at the University of Maryland Electron Ring (UMER), a very low-energy, high-current storage ring which has been in operation for about 20 years. In the first section, we describe the most salient features of UMER including its lattice and injection. In section 1.2, we review the basic physics of emittance versus space charge (SC) dominated beam transport and define the incoherent and coherent betatron tune shifts. Section 1.3 introduces the concept of closed orbit (CO) and its special features in UMER. In section 1.4, we discuss resonances and effects from space charge. In section 1.5, we cover soliton trains and a multi-stream instability, and summarize in section 1.6 research in nonlinear optics in UMER. In section 1.7, we present our personal view of the UMER project, its successes, problems and prospects. To conclude, we present in section 1.8 on computer resources an example, using the code WINAGILE, of an ideal closed orbit in UMER and correction when magnet errors are present. The computer resources include Mathcad worksheets for all basic calculations in UMER, as well as MAD-X and WINAGILE input lattice files.

1.1 UMER lattice and injection

UMER's original motivation was to model beam transport in a recirculator for heavy-ion inertial fusion. Beam transport in such a hypothetical machine entails beam physics phenomena in regimes of high space charge (SC) intensity, both in transverse and longitudinal dynamics [1, 2]. UMER, completed in 2004 [3, 4], except for the extraction section, employs an alternating-gradient focusing lattice over a 36-sided polygon inscribed in an 11.52 m circumference. In this section, we describe in detail UMER's lattice and injection. A distinctive feature of the lattice is the high density of magnets, which allows containing beams from the emittance-dominated to the high SC intensity regimes without changing external focusing (see next section).

As illustrated in figure 1.1, the main lattice consists of 18 focusing/bending sections, each containing two FODO cells (four quadrupoles) and two bending dipoles. These air-core elements are based on printed circuits (PC) which provide the low gradient or fields necessary to focus/bend the low energy electron beam. (The design and measurement of the magnets is discussed in [5].) The chamber at the center of each ring section ('RC' in figure 1.1) houses a fast beam-position monitor (BPM) and fluorescent screen diagnostics. Three of the ring sections contain glass adapters for induction modules or wall-current monitors. Single-turn injection of the 10 keV, 20–100 ns electron beam, with a 60 Hz repetition rate, is set up in a direction such that the Earth's magnetic field provides 1/3 of the bending, approximately. While the Earth's field helps to bend the beam smoothly around the ring, it also complicates beam alignment, especially on injection. Helmholtz coils for the Earth's field compensation are used over the straight part of the matching/injection section as well as along the ring chambers, the latter coils for minimizing the vertical

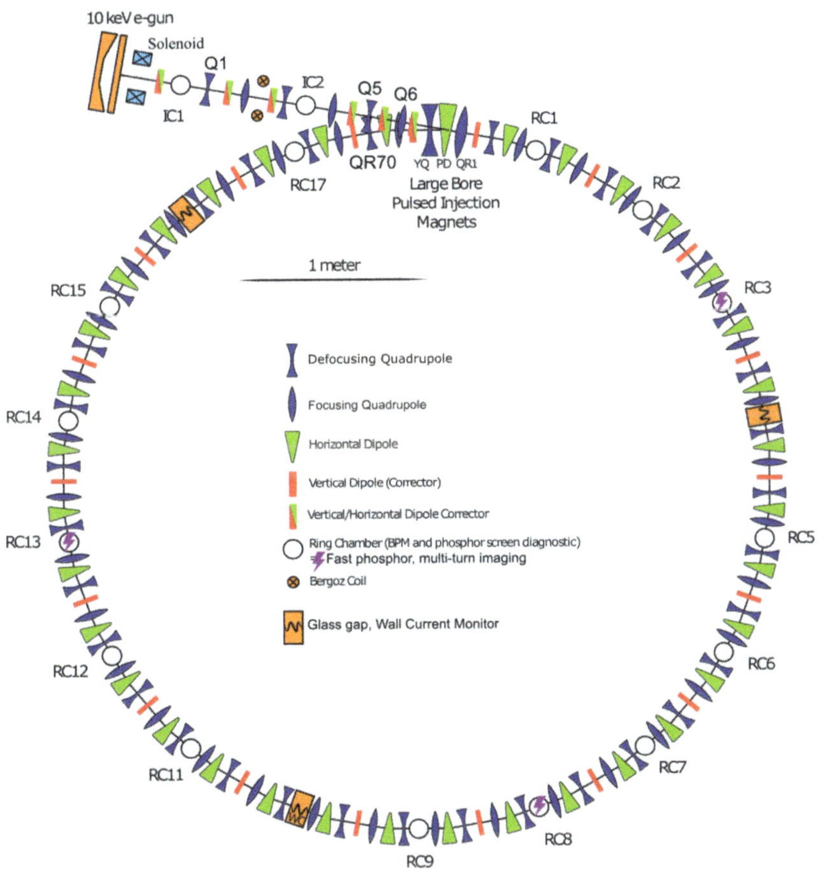

Figure 1.1. Schematics of University of Maryland Electron Ring (UMER). A typical beam bunch is five meters long. Adapted from L Dovlatyan PhD thesis [6], by permission.

Figure 1.2. Photograph of the 11.5 m circumference UMER.

deflection caused by the Earth's field horizontal component. Figure 1.2 shows a photograph of UMER. The red box on the upper left corner hosts the electron gun.

The pulsed injector, also shown in some detail in figure 1.1, consists of two large air-core magnetic PC quadrupoles, a wire-wound dipole and a number of steering elements (only one is shown in figure 1.1). The quadrupole labeled YQ is rotated (yaw angle) 20° relative to Q5–6; in this way, YQ acts as both Q7 and QR72. The injection pulsed dipole (PD), on the other hand, is centered at the intersection of three lines: matching line (Q6-PD), RC1 ring line (PD-QR1), and RC18 line (QR71-PD). Injection and steering are accomplished such that quadrupole YQ can act as a combined-function element, i.e., for focusing and bending into the ring lattice. The injection dipole, PD, is set up with a bending field that switches polarity before the beam completes one turn. The polarity swing is asymmetrical because the bending field opposes the action of the Earth's field on injection but adds to it for recirculation.

1.2 Emittance, space charge and betatron tune shifts

We summarize in this section material from chapter 4 of [7]. For simplicity, we will assume that we have an axisymmetric beam transport problem. In this model, the evolution of the effective beam radius $R(s) \equiv 2X(s) = 2\sqrt{\langle x^2 \rangle}$ is given by the differential equation

$$R''(s) + \left[k_0^2(s) - \frac{K}{R^2(s)}\right]R(s) - \frac{\tilde{\varepsilon}_x^2}{R^3(s)} = 0, \tag{1.1}$$

The angular brackets in the definition of R(s) indicate average over phase space in the (x, x') plane [7]. The double prime indicates second derivative with respect to the transport distance 's'. External focusing is represented by the function $k_0^2(s)$, the *defocusing* effect of linear space charge by $-K/R^2(s)$, and the *natural transverse spread* of the beam by the last term with the *4 × rms* or *effective emittance* $\tilde{\varepsilon}_x = 4\varepsilon_{xrms}$. The parameter K is called the *generalized beam perveance* and is equal to

$$K = \frac{I}{I_0}\frac{2}{(\beta\gamma)^3}. \tag{1.2}$$

I is the beam current, I_0 is a characteristic current equal to 17 kA, approximately, for electrons, and β, γ the standard parameters from special relativity. In the *uniform-focusing*, or smooth, approximation, we set the second derivative term in equation (1.1) to zero and represent focusing by the constant k_0. Then, we write

$$\left[k_0^2 - \frac{K}{a^2}\right]a - \frac{\tilde{\varepsilon}_x^2}{a^3} = 0. \tag{1.3}$$

We consider two limits: the zero current ($K = 0$), non-zero emittance limit, and the zero emittance, non-zero current limit. In the first case we obtain for the beam radius

$$a_{I\to 0} \equiv a_0 = \sqrt{\frac{\tilde{\varepsilon}_x}{k_0}}, \tag{1.4}$$

while the second limit yields

$$a_{\tilde{\varepsilon}_x\to 0} \equiv a_B = \frac{\sqrt{K}}{k_0}. \tag{1.5}$$

The uniform-focusing model can be related to a *periodic* lattice with full-lattice period S by defining the zero-current or 'bare' phase advance per period $\sigma_0 = k_0 S$. Furthermore, if the lattice has N cells on a circumference of radius ρ_m, and λ_0 is the betatron wavelength, the *bare tune* is the number of betatron oscillations *per turn*, or

$$\nu_0 = \frac{2\pi\rho_m}{\lambda_0} = k_0\rho_m. \tag{1.6}$$

By the same token, linear space charge changes the tune to

$$\nu = \frac{2\pi\rho_m}{\lambda} = k\rho_m,\ k^2 = k_0^2 - \frac{K}{a^2}, \tag{1.7}$$

from equation (1.3). For *weak space charge*, $K/k_0^2a^2 \ll 1$ and we write for the *incoherent betatron tune shift*

$$\Delta\nu_{\text{inc}} = \nu_0 - \nu \cong \frac{K\rho_m}{2k_0 a^2} \cong \frac{K\rho_m}{2\tilde{\varepsilon}_x}, \tag{1.8}$$

where we have used equation (1.4). (Note that many authors define the tune shift as ν—ν_0, which is negative.)

The effect of direct SC can be quantified also with the dimensionless *tune depression*, defined as

$$\eta = \frac{k}{k_0} = \left(\frac{a_0}{a}\right)^2, \tag{1.9}$$

where we have used $\tilde{\varepsilon}_x = a_0^2 k_0 = a^2\, k$. The range of η is from 1 for zero current to zero when external focusing exactly balances SC (equation (1.7)). In the latter case $k = 0$, which entails zero emittance, i.e., perfectly laminar flow with $\lambda \to \infty$.

Another important effect on beam transport is caused by image forces on off-centered beams. For small beams, relative to the vacuum pipe radius, and small off-center displacements the *coherent tune shift* is given by [7]

$$\Delta\nu_{\text{coh}} = \nu_0 - \nu_{\text{eff}} \cong \frac{K\rho_m}{2\tilde{\varepsilon}_x}\left(\frac{a_0}{h}\right)^2, \tag{1.10}$$

where 'h' is the pipe height.

As mentioned above, the high density of quadrupoles in the UMER lattice allows us to transport all beam currents without changing the bare tune. This can be understood from the following approximate equation for the average beam radius in a uniform-focusing approximation of the lattice (see equations (1.4) and (1.5)):

$$a \cong \left(a_B^2 + a_0^2\right)^{1/2} = \frac{S}{\sigma_0}\left(K + \tilde{\varepsilon}_x \frac{\sigma_0}{S}\right)^{1/2}, \tag{1.11}$$

where $S = 0.32$ m is the full-lattice period. The beam radius 'a' is proportional to $S^{1/2}$ in the regime of emittance-dominated transport, and to S in the limit of strong space charge. Thus, having closely spaced focusing elements is more efficient for the transport of intense beams, other parameters being equal, than for the transport of low current.

Alternative lattices with one-half ($S = 0.64$ m) or even one-fourth ($S = 1.28$ m) the number of quadrupoles are in principle possible in UMER. Clearly, for given beam current and emittance, the beam radius is larger for transport in these lattices, corresponding to particle motion that is more laminar, i.e., more SC dominated if the emittance is kept the same. Furthermore, injection without the tilted quadrupole YQ is feasible in alternative lattices. An additional lattice scheme with a three-fold symmetry had been used for nonlinear optics studies, as described in section 1.6.

Table 1.1 summarizes the main parameters of UMER. The betatron tunes are determined experimentally [8]. For comparison, a simple-minded matrix calculation without bending dipoles yields equal bare tunes $\nu_{x0} = \nu_{y0} = 6.83$ (equations 3.16, 3.17 in [7]). If the edge effects of bending dipoles are included, we obtain, using the code ELEGANT (see appendix A in [7]), $\nu_{x0} = 6.82$, $\nu_{y0} = 6.88$. These edge effects comprise not only the standard vertical edge focusing from the rectangular bending magnets (see, e.g., chapter 3 in [7]) but also the field variation at the fringes. An additional effect from the Earth's magnetic field in the low-energy UMER can be

Table 1.1. Main parameters of the University of Maryland Electron Ring (UMER).

Energy, $\boldsymbol{E}$	10 keV
Super-period, $\boldsymbol{N}$	1
Circumference	11.52 m
Beam current, $\boldsymbol{I_B}$	0.06–100 mA
Horizontal betatron tune, $\boldsymbol{\nu_{x0}}$	6.68
Vertical betatron tune, $\boldsymbol{\nu_{y0}}$	6.85
Momentum compaction, $\boldsymbol{\alpha}$	0.0214*
Tune depression, $\boldsymbol{\nu/\nu_{x0}}$	0.15–0.86
Coherent tune shift, $\Delta\nu_{coh}$	0.00036–0.61
Emittance (rms, norm.), $\tilde{\varepsilon}_n$	<3.2 μm
Natural hor. chromaticity, $\boldsymbol{\xi_X}$	−7.31*
Natural vert. chromaticity, $\boldsymbol{\xi_Y}$	−7.39*

*Elegant, with no space charge.

modeled in ELEGANT through the 'FSE' ('field strength error') parameter. In effect, ELEGANT keeps the reference orbit for 36, 10° bending dipoles, while using actual 7°, approximately, dipoles.

The most interesting basic calculations in UMER relate to the effects of incoherent and coherent SC. We limit our brief discussion here to transverse dynamics effects, although longitudinal space-charge phenomena also play a significant role in UMER. At the nominal beam current of 0.6 mA, beam transport in UMER is *emittance-dominated*, although the fractional incoherent SC tune shift, $\Delta\nu_{inc}/\nu_{x0}$, equation (1.8), is sizable, about 14%. The corresponding SC *tune depression* is $\nu/\nu_{x0} = 0.86$. At the other end of the beam current range, 100 mA, we obtain a fractional incoherent SC tune shift of 85% (!) or an SC tune depression of $\nu/\nu_{x0} = 0.15$. UMER operates most of the time, however, with 6.0 mA beam current; the tune depression in this case is $\nu/\nu_{x0} = 0.64$, which corresponds to strong SC dominated transport.

The *coherent tune shift* caused by image forces is also appreciable in UMER, despite the low energy. We obtain, from equation (1.10), $\Delta\nu_{coh} = 0.0035$ for 0.6 mA. The more general equation (4.55) in [7] yields $\Delta\nu_{coh} = 0.61$ for 100 mA, and 0.035 at the typical 6.0 mA operating current.

1.3 Closed orbit

The beam centroid displacement in the radial direction as measured by a beam-position monitor (BPM) at a given location s in a ring lattice includes contributions from the CO (also called 'closed orbit distortion'), betatron function, dispersion, and BPM error:

$$x_{BPM}(s) = x_{c.o.}(s) + x_{\beta}(s) + x_{\delta}(s) + \varepsilon_{BPM}. \tag{1.12}$$

The second and third terms above comprise the displacement $x(s) = x_{\beta}(s) + D_{0x}(s)\delta$, which is the sum of the homogeneous and particular solutions of the equation

$$x''(s) + \kappa_x(s)x = \frac{1}{\rho(s)}\delta. \tag{1.13}$$

We have $\kappa_x(s) \equiv k_0^2(s)$ (see equation (1.1)), $x_\delta = D_{0x}(s)\delta$ for linear dispersion D_{0x} without space charge, δ is the fractional momentum error, and $\rho(s)$ is the *local* orbit radius. The last term in equation (1.12) denotes a BPM error from calibration and other sources. With no magnet, momentum or BPM errors, the BPM would directly measure the betatron oscillations whose amplitude is proportional to $\beta_x^{1/2}(s)$. Because the betatron term is $x_\beta(s) = x_{\mathrm{BPM}}(s) - x_{c.o.}(s)$, with magnet but no momentum errors, the *c.o.* defines the *reference (or equilibrium) orbit* for particles of design momentum p_0.

Equilibrium orbits in UMER are determined by the ambient *B*-field (Earth's *B*-field plus effects from ferromagnetic structures), the strengths of bending dipoles, and by the unavoidable dipole fields from off-axis motion through the quadrupoles. Ideally, the main bending dipoles would be powered with the same current, 2.35 A, in the presence of a −0.40 G uniform vertical component of the Earth's *B*-field. However, because of the non-uniform ambient *B*-field and mechanical imperfections, the dipoles are set individually with currents in the 2.0–2.6 A range, approximately.

Figure 1.3 illustrates three examples of ideal equilibrium orbits over three UMER sections, i.e., six FODO periods, for three current settings of the bending PC dipoles. The vertical component of the Earth's *B*-field is modeled in the code WINAGILE (see appendix A of [7]) with 60 small negative kicks (of the order of −1 mrad) over the length of the ring section (0.64 m). Further, the ring dipoles are set to 10° bending to define the reference trajectory, but they are divided into four rectangular bends over which positive kicks act to 'undo' the action of Earth's field over half a ring section; the strength of these positive kicks will depend on the actual current fed to the PC dipoles. For the orbit closest to the quadrupole magnetic axes, an offset of 0.6 mm (and essentially zero slope) is required between quadrupoles, i.e., at the bending dipoles and at the ring diagnostics chambers (see figure 1.1). Notice also the small horizontal offset of the centroid orbit through the quadrupoles. Naturally, different injection settings, i.e., values of (x_0, x_0'), are required to obtain the orbits

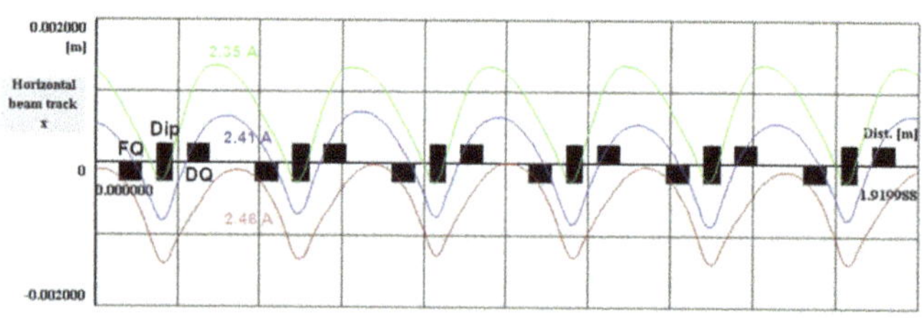

Figure 1.3. Three ideal horizontal (*x*-component) equilibrium orbits over six FODO periods in UMER calculated with WINAGILE. The bending dipole currents are 2.35, 2.41 and 2.46 A, and the ambient vertical *B*-field is − 0.4 G. FQ/DQ: focusing/defocusing quadruple; DIP dipole. Reproduced from [9] with minor changes, with the permission of AIP Publishing.

shown in figure 1.3. Additional closed orbit calculations, which include magnet errors, are presented at the end in the section on computer resources.

In practice, the machine is tuned to obtain a first-turn orbit going as close as possible through the quadrupole centers. Subsequent turns are then optimized to minimize centroid oscillations.

The vertical component of the centroid orbit is minimized by using 18 short PC steering correctors, one per ring section, as well as Helmholtz coils designed to compensate for the radial component of the Earth's field. A major factor that affects the closed orbits, especially for large beams in alternative lattices ($S = 0.64$, 1.28 m) are image forces, which add to demands for improved injection and steering. Details of the numerical and simulation procedures for tuning UMER are discussed in chapter 10 and appendix K of [10].

1.4 Betatron resonances and space charge

In the coherent resonance theory [11], we have $n(\nu_0 - \Delta\nu_{\text{eff}}) = p$, ($n$, p integers) for the condition of betatron resonances of an isotropic beam. The effective tune shift $\Delta\nu_{\text{eff}}$ is related to the direct SC tune shift $\Delta\nu = \nu_0 - \nu$ through a *coherent mode* coefficient C, such that $\Delta\nu_{\text{eff}} = C\Delta\nu$. Thus,

$$n(\nu_0 - C\Delta\nu) = p. \tag{1.14}$$

The coherent mode coefficient is applicable to linear resonances only, i.e., $n = 1, 2$. For $n = 1$, corresponding to a dipole resonance, $C = 0$: the condition for integer resonances is not changed by space charge [11]. When $n = 2$ in equation (1.14), on the other hand, we have a quadrupole, i.e., half-integer resonance. In this case, $C = ½, ¾$, corresponding to symmetric and antisymmetric envelope modes, respectively, discussed in section 7.1 of [7] in the context of a mismatched K–V beam. With direct SC, resonance occurs when the *mode tune* equals the harmonic integer 'p'. The mode tunes are given by [7]

$$\begin{aligned} \nu_{\text{sym.}} &= \sqrt{2\nu_0^2 + 2\nu^2}, \text{ symmetric mode,} \\ \nu_{\text{asym.}} &= \sqrt{\nu_0^2 + 3\nu^2}, \text{ antisymmetric mode.} \end{aligned} \tag{1.15}$$

As shown in figure 1.4, curves of ν_0 versus ν for $p = 10, 11, 12, 13$ illustrate second-order resonances for the two envelope modes [12]. The depressed tune ν is calculated using the smooth approximation expression [13]

$$\nu = \nu_0\left[\sqrt{1 + u^2} - u\right], \ u = \frac{KL_C}{4\pi\nu_0\tilde{\varepsilon}_x}. \tag{1.16}$$

The dimensionless parameter 'u' in the last equation is defined in terms of the beam perveance K (equation (1.2)), the machine circumference L_C, the bare tune ν_0, and the effective emittance $\tilde{\varepsilon}_x = 4\varepsilon_{xrms}$.

The curves in figure 1.4 are applicable to a 6.0 mA electron beam at 10 keV, with 26.2 μm effective emittance in the standard ($S = 0.32$ m) UMER lattice. It is important to realize that we are not plotting equation (1.16) but rather expressions

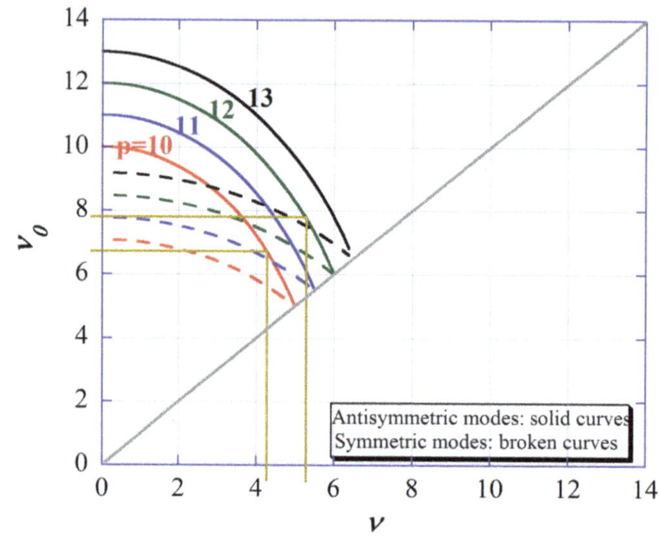

Figure 1.4. Mode resonance curves for a 6.0 mA, 10 keV, 26.2 μm (effective emittance) beam in UMER. Two examples of resonances are shown at $p = 10$ and 12. See text and [12].

of the form $\sqrt{-3\nu^2 + p^2}$ (antisymmetric modes), and $2^{-1/2}\sqrt{-2\nu^2 + p^2}$ (symmetric modes). However, (ν, ν_0) must also satisfy equation (1.16) at mode resonances. For example, for $p = 10$, an antisymmetric mode resonance occurs for $\nu_0 = 6.72$, for which $\nu = 4.28$. i.e., with a tune shift $\Delta\nu = \nu_0 - \nu = 2.4$. Another antisymmetric resonance occurs for $p = 12$, at $\nu_0 = 7.80$, for which $\nu = 5.27$, and $\Delta\nu = 2.5$. See also the last section on computer resources.

We reported in [14] calculations of the incoherent tune shift based on the displacement of a half-integer horizontal resonance band of a 10 keV, 6.0 mA electron beam, as observed in a chart of beam losses versus estimated tunes (figure 7.4 in [14]). For the analysis, we had used the coherence resonance theory discussed above. (The two numerical examples just mentioned—and illustrated in figure 1.4—are in the ballpark of the analysis in [14], which was based on the coherent mode coefficients.) However, additional studies by Ruisard and Dovlatyan [6, 10] did not display the exact same resonance band structures seen before with 6.0 mA (10th turn). Moreover, these latter investigations did not reveal a clear-cut difference in resonance structure between emittance and SC dominated beams. Differences in CO tuning, envelope matching and injection, and longitudinal expansion from SC for various beam currents made the comparison inconclusive at best.

Dovlatyan studied the resonances for three beam currents, 0.6, 6.0 and 20 mA, all at 10 keV. We show below in figures 1.5(a) and (b) the charts of beam losses versus estimated tunes (at the 10th turn) for 0.6 and 6.0 mA. The beam loss is the fraction of beam that survives relative to the injected beam as measured with a wall-current monitor. Unlike the results reported in [14], no horizontal half-integer resonances are seen in the new experiments at the same 10th turn; instead, a new vertical

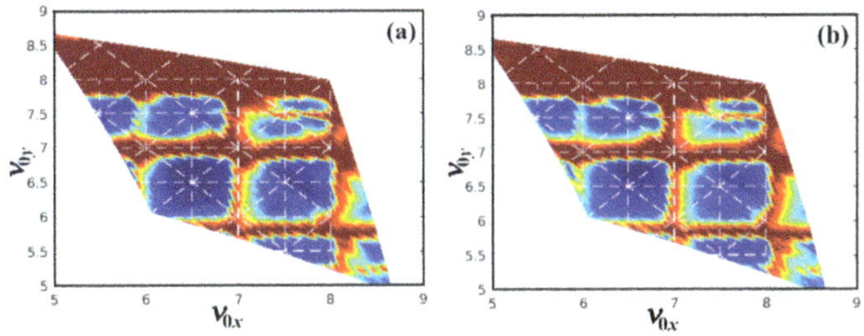

Figure 1.5. Beam losses at 10th turn as functions of estimated bare betatron tunes for (a) 0.6 mA, and (b) 6.0 mA. Dark red indicates 100% beam loss and blue 100% beam transmission. Adapted from L Dovlatyan PhD thesis [6], by permission.

half-integer resonance is observed at $\nu_{0Y} = 7.5$. The differences can be ascribed to improvements in beam steering and matching (which can delay the development of the half-integer resonances), and refined tune measurements. In fact, additional studies by Dovlatyan using NAFF (Numerical Analysis of Fundamental Frequencies) techniques [15, 16] to measure tunes reveal the appearance of both horizontal and vertical half-integer resonances for the 0.6 mA beam (25th turn) and a half-integer horizontal resonance for the 6.0 beam (15th turn). The different rates of longitudinal expansion (debunching) from space charge (section 6.5 in [7]), as well as the emittance evolution, for the different beam currents complicate the comparison of resonance band structures. The conclusion of these studies is that the integer resonances are not affected by direct space charge, as predicted by theory [11], but the expected displacement of the half-integer bands is difficult to ascertain. Additional considerations are given when we discuss the experiments in Paul traps in chapter 5.

1.5 Soliton trains and multi-stream instability

The first observation of soliton wave trains in electron beams was made in UMER in 2012 [17]. Solitons can form in water, EM radiation, plasmas and other media when the steepening of large wave perturbations are balanced by the wave dispersion [18]. With a large density perturbation applied at the tail of a long bunch by means of photoemission at the cathode, a fast wave develops into a soliton wave train after 2–3 turns. The pulses maintain their shapes and other properties of solitons. The observations have been successfully simulated with the WARP code (see appendix A in [7]). Figure 1.6 shows the formation and evolution of a soliton wave train on a 22 mA, 10 keV beam with a 25% perturbation. The initial beam (bottom trace) is measured with the Bergoz transformer before injection into the ring, while the rest of the traces are from the wall-current (WC) monitor in RC10 (figure 1.1).

Without longitudinal focusing, the beam in UMER completely debunches, i.e., turns into a DC beam that fills the entire ring (see end of section 6.5 in [7]). The WC monitor is blind to the DC beam, but partial re-bunching and later the onset of an instability are observed as the beam overlaps itself around the ring. For the 6.0 mA, 10 keV beam, partial re-bunching happens after some 40 turns, and an instability

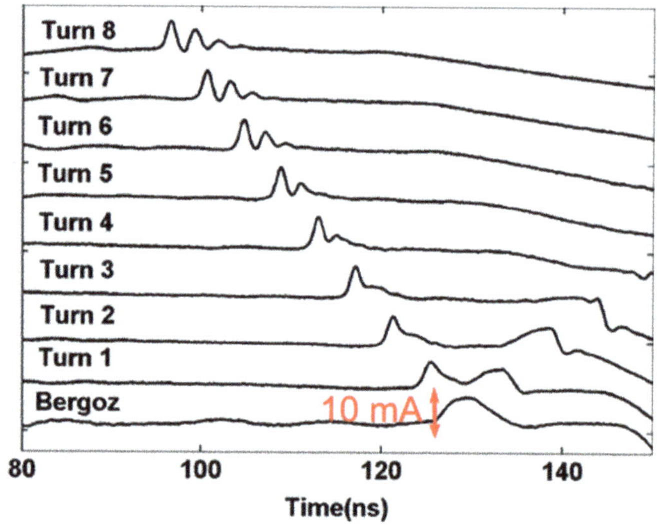

Figure 1.6. Injected and turn-by-turn beam current of 22 mA beam with a 25% perturbation. The fast wave on the trail end of a 100 ns bunch moves to the left and evolves into a soliton wave train. Adapted with permission from [17], copyright (2013) by the American Physical Society.

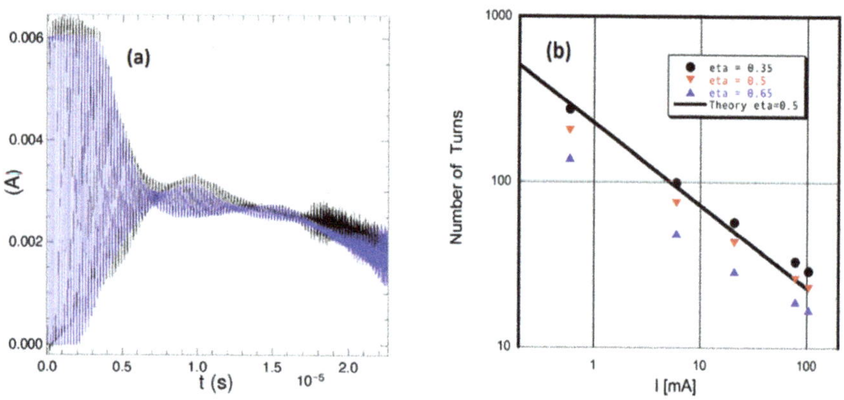

Figure 1.7. (a) Current evolution for a (initially) 6.0 mA, 10 keV beam in UMER. The oscillation amplitude decreases as the beam expands and overlaps itself. Re-bunching and the sudden onset of an instability are observed. (b) Number of turns for instability onset vs. beam current: theory (solid line) and WARP simulations (symbols) for three initial fill ratios. Reproduced from [19], with permission from AIP Publishing.

develops suddenly after another 40 turns, or at 17 μs from injection. Figure 1.7(a) shows the beam current versus time as measured with the WC monitor in RC10 (see figure 1.1). Filamentation of the beam in longitudinal phase space occurs as the charge wraps around the ring. When the separation in velocity space between filaments is comparable to the sound speed c_s in the beam, the filaments interact, and the instability develops. There is good agreement among theory, simulations with WARP, and experiment for different line-charge densities at the center of the bunch [19]. The theory is based on one-dimensional plasma waves with a dispersion relation, modified by the presence of the conducting wall, for perturbations in the long wavelength limit, i.e., for wavelengths large compared to the vacuum pipe radius [13]. Assuming an initially rectangular current profile, the bunch ends expand with a velocity $2c_s$, while a rarefaction wave propagates with the velocity c_s towards the bunch center. An expression for c_s in terms of the peak line-charge density of the beam and a geometry factor g was given in equation (6.57) of [7]. The long-term evolution of the beam density and velocity axial profiles is modeled with a 'similarity solution' whereby the velocity is a linear function of distance with a time-dependent slope [19]. The density and velocity profiles are connected through the cold fluid continuity and force balance equations. Thus, the density versus distance is derived from the similarity solution when combined with results from WARP simulations for the longitudinal phase space. Furthermore, a dispersion relation is constructed for waves in a multi-stream beam with a correction term that limits the maximum growth rate of perturbations. It is found that instability occurs earlier, i.e., at a smaller number of turns, for beams whose initial bunch lengths L_i are larger relative to the ring circumference L_c. Figure 1.7(b) illustrates this for three fill ratios $\eta = L_i/L_c$. Beam losses were taken into account in simulations of the 6.0 mA beam; the effect of these losses was to delay the onset of the instability.

1.6 Nonlinear optics

The onset of nonlinear betatron resonances, i.e., those of third or higher order, depend on the particle's amplitude (see chapter 7 in [7]). Therefore, the fields from nonlinear magnets (e.g., sextupoles or octupoles—see appendix B in [7]) or from errors of nominally linear magnets can lead to significant reduction of *dynamic aperture*, DA, defined as the beam size at the limit of stable orbits, at working tunes close to or at the single-particle resonance condition

$$n_x\nu_{x0} \pm n_y\nu_{y0} = Np, \tag{1.17}$$

where n_x, n_y, p and N are integers, and ν_{x0} and ν_{y0} are the horizontal and vertical bare tunes (the notation Q_{x0}, Q_{y0} is used in Europe).

In addition, the driving nonlinear fields induce betatron tunes that also depend on amplitude, leading to tune spreads. However, under certain conditions, to be discussed in some detail in section 3.3, the tune spreads induced by nonlinear magnets can mitigate nonlinear resonances without reduction in DA. The flexibility of the UMER lattice has facilitated the investigation of these effects in detailed particle-in-cell (PIC) simulations and preliminary experiments with low space charge [10, 20].

A modified UMER lattice with ideal three-fold symmetry (3.84 m super-period) was designed to accommodate a 25 cm nonlinear magnet insert between two dipoles in a 20° section; the lattice is shown in figure 1.8(a). The insert consists of seven short PC octupoles. The overlap of the octupole fields is tailored to follow closely the required profile to satisfy *quasi-integrable* (QI) conditions in combination with the linear optics of the rest of the ring: the beam must be round in the nonlinear insert, while the phase advance in the rest of the linear ring must be an integer multiple of 180°. These conditions allow for 'chaotic but bounded' orbits with moderate tune

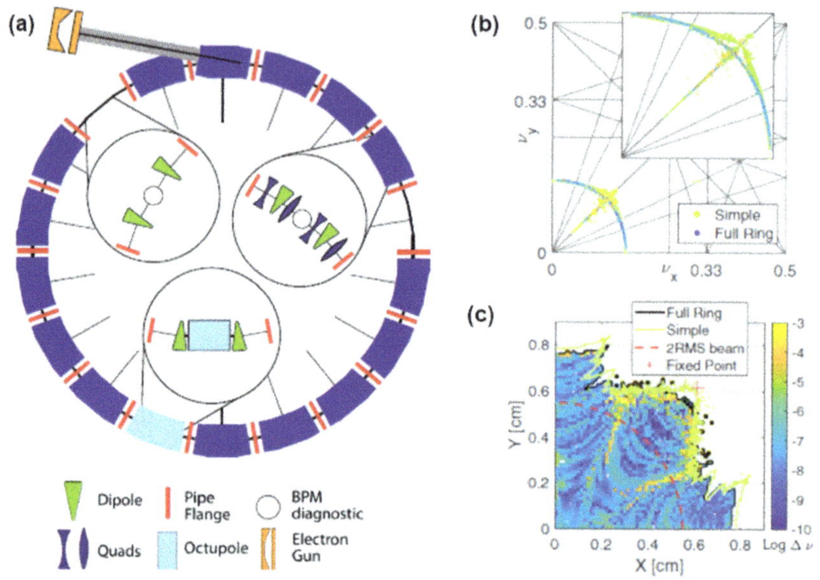

Figure 1.8. (a) Modified UMER lattice for QI studies with octupole insert. WARP simulations after 1024 turns in simple and full ring lattices: (b) Tune footprint and (c) Representation of DA and resonant structure. Reproduced with permission from [20], copyright (2019) by the American Physical Society.

spreads, unlike the fully integrable systems whereby non-chaotic orbits can be maintained with large tune spreads (section 3.3). Naturally, space charge and chromaticity effects cause deviations from the desired conditions.

The beams for the PIC simulations and initial experiments have low SC intensity parameters, corresponding to incoherent tune shifts of 0.005–0.3 at horizontal and vertical bare tunes of 6.8 in the standard UMER lattice. The operating tunes in the modified lattice with the octupole insert, however, are close to 3.13, which leads, according to WARP simulations, to a tune spread of 0.13. Figures 1.8(b) and (c) show results of WARP simulations after 1024 turns in a simple model (octupole insert plus single focusing kick representing the rest of the UMER lattice), and in the insert plus full lattice, for a peak octupole strength of 50 T m^{-3}. The tune footprint in figure 1.8(b) displays the 0.13 tune spread, and figure 1.8(c) is a representation of the DA for initial conditions in configuration space spanned by the range of X, Y values shown. In addition, the resonant structure is represented by the different colors, with blue indicating the most regular orbits, and yellow the more chaotic ones. See also [16].

The conclusion of these studies is that very tight control of orbit distortion in the octupole section (<0.2 mm) is required to preserve a sizable DA, a condition not achieved yet in the experiments. First-turn orbit measurements over the location of the nonlinear insert reveal centroid displacements of up to 0.69 mm in the horizontal plane, and 0.10 mm in the vertical plane. Improvements may require better ambient-field compensation, optimized injection, and dispersion matching, among other measures. Once UMER overcomes these difficulties, it will be in a position to test nonlinear dynamics with significant space charge.

1.7 Prospects

The UMER project was the logical successor to research in a 5 m long, 36-solenoid channel, that was operational until around 1992. However, a circular lattice with strong focusing involved new challenges that were not always anticipated. One of these was the effect of the ambient *B*-field on a 10 keV electron beam, particularly on injection and the closed orbit. Although several thousand turns were achieved at low current with all 72 quadrupoles, it was not possible to do the same with the alternative lattice involving 36 quadrupoles and without the tilted wide-aperture injection quadrupole (so-called Collins injection). Part of the problem was the lack of diagnostics such as BPMs in the injection *Y*-section (see figure 1.1). Thus, in our view, a modified UMER should involve shielding of the Earth's *B*-field, as well as a new electrostatic injection system with diagnostics and a septum-type of injector to avoid injection through a quadrupole edge. (The new geometry, in principle, would also improve the periodicity. This is a lesson taken from EMMA—see chapter 4.) In addition, the extraction could be combined with injection in an X-shape geometry. Although an extraction section was designed and simulated [22], it was deemed to have similar problems to the injection section. Thus, we decided to implement a 'poor man's' extraction that allowed us to image small beams by the turn, and optimally at 6 mA (nominal current). A fast 'knockout' pulse was applied to the vertical plates of the BPM at RC6 (see figure 1.1) and the beam was detected downstream at a fast screen partially inserted into the vacuum pipe at RC8. The timing of the pulse was selected for the desired turn number, and a gated PIMAX II camera with a 3 ns resolution was used to collect images. The scheme is illustrated in figure 1.9. Two additional 'extraction' locations were planned for *by-the-turn* emittance measurements.

Furthermore, matching of the dispersion function could be implemented by adding a bending dipole to the injection system. We have calculated that mismatched dispersion can be detrimental, at a 10%–20% level relative to the effect of direct SC alone for a 1% (200 eV) uncorrelated energy spread [7, 21]. Lastly, chromaticity correction via sextupole magnets is feasible; a PC sextupole design was completed and schemes for corrections were underway [23].

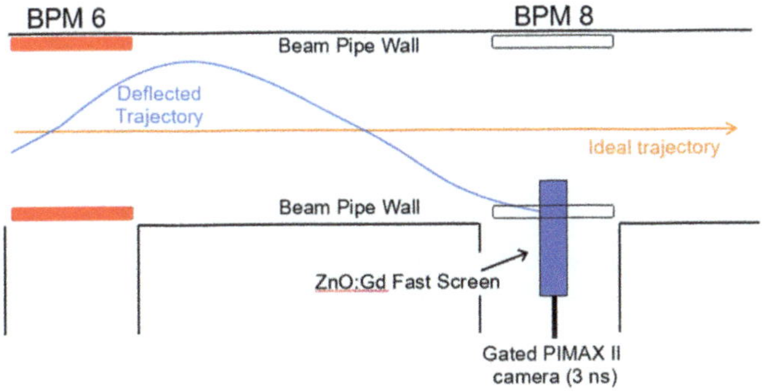

Figure 1.9. Schematics of 'poor man's' extraction system in UMER. Fast knockout deflection of beam at beam-position monitor 6 (RC6 in figure 1.1) and detection at fast screen in RC8.

We will not speculate on the chances of any of these ideas being implemented in the near future. We simply offer the information with possible similar 'budget' projects in mind at U. of Maryland or elsewhere. We can only assure that UMER will continue playing an important role in educating future accelerator physicists and engineers.

1.8 Computer resources

The author's website contains the Mathcad file **Umer2019-Simplified-v2.xmcd** that reviews many single-particle and basic space charge calculations for UMER. Furthermore, the file **Dipole Review.xmcd** includes calculations related to the closed orbit curves in figure 1.3. The resonance curves shown in figure 1.4 can be reproduced with the Mathcad script **Mode_Res_Curve.xmcd**.

In appendix B, we illustrate the use of the code WINAGILE with a simple example. Here, we present a more complex application. The lattice file **ring-ex2.lat** contains an idealized model of the UMER machine with 36 bending dipoles and 72 quadrupoles (see figure 1.1) and a constant external B-field of −0.4 G (pointing downwards). The latter is approximated in the kick file **wina-ex3.kck** which specifies constant kicks of 1–2 mrad over drifts of about 1–2 cm. In addition, there is a constant kick (6.1 mrad) applied over the effective length of each quadrupole, and positive vertical kicks (18 mrad) applied at each SBEND element (the bending dipoles are represented by two sector, 5°-bends and a zero-length *corrector* magnet between them.) The effect of the positive kicks is to simulate the reduced field of the dipoles that is required in the presence of the constant −0.4 G Earth's field assumed. The magnet *random errors* implemented are $\sigma = 0.2$ mrad, bending dipole kick error; $\sigma = 2.0$ mrad, transverse tilt error of dipoles; and $\sigma = 0.2$ mm, transverse shift error of quadrupoles. The error distribution is truncated at 2.2σ.

Figure 1.10 displays three horizontal COs over one turn (11.52 m or 36 FODO periods): the ideal CO (with the constant Earth's field), the CO with magnet errors, and the corrected CO. Twenty dipole correctors are employed along with 76 monitors (all quadrupoles are used as virtual BPMs.)

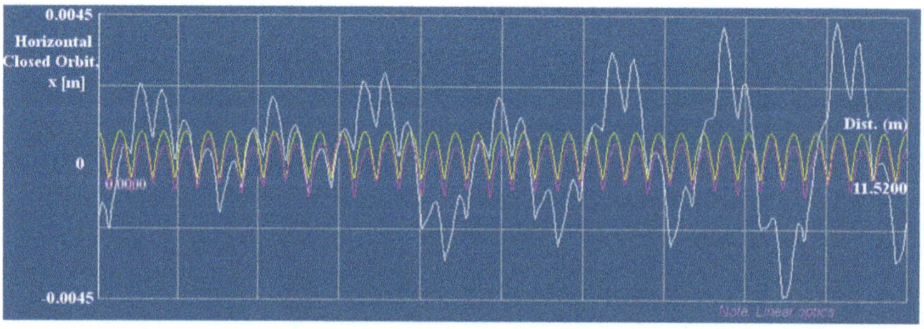

Figure 1.10. WINAGILE three horizontal COs in ideal model of UMER (figure 1.1). In all three cases, there is a constant vertical B-field equal to −0.4 G. The yellow trace represents the CO in the absence of magnet strength and alignment errors. The white trace is the CO in the presence of errors (see text), and the purple trace is the corrected CO. The WINAGILE files required to reproduce these results are available online.

References

[1] Reiser M 1996 Physics of intense charged-particle beams for heavy-ion inertial fusion *Proc. Int. Symp. on Heavy Ion Inertial Fusion* 32–33 *(Princeton NJ, Sept. 6–9)* (Fusion Engineering and Design) 133–40

[2] Reiser M, O'Shea P G, Kishek R A and Bernal S *et al* 1999 The Maryland electron ring for investigating space-charge dominated beams in a circular FODO system *Proc. of the 1999 Particle Accelerator Conf. (New York)* 234–6

[3] Bernal S *et al* 2005 Commissioning of the University of Maryland electron ring (UMER) *Proc. of PAC (Knoxville, TN)* (FOAD005)

[4] Bernal S, Bai G, Beaudoin B and Feldman D *et al* 2006 Space-charge beam physics research at the University of Maryland Electron Ring (UMER) *Proc. of 39th ICFA Advanced Beam Dynamics Workshop, HB2006 (Tsukuba, Japan)* 218–22

[5] Zhang W W, Bernal S, Li H, Godlove T, Kishek R A, O'Shea P G, Reiser M and Yun V 2000 Design and field measurements of printed-circuit quadrupoles and dipoles *Phys. Rev. ST Accel. Beams* **3** 122401

[6] Dovlatyan L 2020 Study and mitigation of transverse resonances with space charge effects at the University of Maryland Electron Ring *PhD Thesis* University of Maryland, College Park

[7] Bernal S 2022 *A Practical Introduction to Beam Physics and Particle Accelerators* 3rd edn (Bristol: IOP Publishing)

[8] Bernal S, Beaudoin B, Cornacchia M and Sutter D 2013 Stability of emittance vs. space-charge dominated beams in an electron recirculator *Proc. of PAC (Pasadena, CA)* (TUPAC31)

[9] Bernal S *et al* 2006 New developments in space-charge beam physics research at the University of Maryland Electron Ring (UMER) *AIP Conf. Proc. 877, 12th Advanced Accelerator Concepts Workshop, AAC06 (Lake Geneva, WI)*

[10] Ruisard K 2018 Design of a nonlinear quasi-integrable lattice for resonance suppression at the University of Maryland Electron Ring *PhD Thesis* University of Maryland, College Park

[11] Baartman R 1998 Betatron resonances with space charge *AIP Conf. Proc.* 448 56–72

[12] Venturini M and Gluckstern R L 2000 Resonance analysis for a space charge dominated beam in a circular lattice *Phys. Rev. ST Accel. Beams* **3** 034203

[13] Reiser M 2008 *Theory and Design of Charged Particle Beams* 2nd (Weinheim: Wiley-VCH)

[14] Bernal S 2018 *A Practical Introduction to Beam Physics and Particle Accelerators* 2nd edn (Bristol: Morgan & Claypool Publishers, IOP Publishing)

[15] Laskar J, Froeschlé C and Celletti A 1992 The measure of chaos by the numerical analysis of the fundamental frequencies. Application to the standard mapping *Phys. D: Nonlinear Phenom.* **56** 253–69

[16] Laskar J 2003 Frequency map analysis and particle accelerators *Proc. of the 2003 Particle Accelerator Conf., PAC (Portland, Oregon, May 12–16)* 378

[17] Mo Y C, Kishek R A, Feldman D, Haber I, Beaudoin B, O'Shea P G and Thangaraj J C T 2013 Experimental observations of soliton wave trains in electron beams *Phys. Rev. Lett.* **110** 084802

[18] Manukure S and Booker T 2021 A short overview of solitons and applications *Partial Differ. Equ. Appl. Math.* **4** 100140

[19] Beaudoin B L, Haber I, Kishek R A, Koeth T W and Antonsen T M 2019 Multi-stream instability of a single long electron bunch in a storage ring *Phys. Plasmas* **26** 052106

[20] Ruisard K, Komkov H B, Beaudoin B, Haber I, Matthew D and Koeth T 2019 Single-invariant nonlinear optics for a small electron recirculator *Phys. Rev. Accel. Beams* **22** 041601

[21] Bernal S, Beaudoin B, Koeth T and O'Shea P G 2011 Smooth approximation model of dispersion with strong space charge for continuous beams *Phys. Rev. ST Accel. Beams* **14** 104202

[22] Ruisard K, Beaudoin B, Bernal S, Butcher J, Sutter D, Haber I, Kishek R A and Koeth T 2013 Design and simulation of an extraction section for the university of maryland electron ring *Proc. of the 2013 Int. Accelerator Conf., IPAC (Shanghai, China)* 1052

[23] Gleason N and Bernal S 2018 *Design of Printed-Circuit Sextupole Magnets for Chromaticity Corrections in UMER* unpublished

Chapter 2

Space charge in the isochronous regime at the Small Isochronous Ring (SIR)

The Small Isochronous Ring (SIR) was originally developed at Michigan State University and is now hosted by the University of Maryland in College Park. It is a strong-focusing machine with fixed bending/focusing fields that can be tuned for *isochronous* operation. The low energy hydrogen ion bunches exhibit a fast longitudinal instability that is driven by coherent transverse space charge. This instability can occur *below* transition energy, unlike the *negative-mass instability* of synchrotrons, which occurs above transition energy. We describe in section 2.1 SIR layout and main parameters. In the next section, we review the momentum compaction, slip factor, transition gamma, and chromaticity. In section 2.3, we discuss briefly the effects of direct space charge (SC) on dispersion, and relate the effect to the fractional incoherent tune change and slip factor. We cover in section 2.4 a derivation of the growth rate of the negative-mass instability; and summarize in section 2.5 results of simulations and some details of experiments. In section 2.6, we give a brief account of prospects for additional research, and conclude in the last section with computer resources.

2.1 Layout and main parameters

The Small Isochronous Ring (SIR) is a low energy, 6.6-m-circumference ion storage ring developed in the early 2000s at Michigan State University in Lansing, MI. Table 2.1 summarizes SIR's main parameters, figure 2.1 shows the top view schematics of the machine, and figure 2.2 is a photograph of the machine.

As shown in figures 2.1 and 2.2, the ring consists of four 90° bend magnets, several dipole magnets and electrostatic quadrupoles, and a Faraday cup and fluorescent screen diagnostics. The Faraday cup provides longitudinal beam profiling with a resolution of 1 ns, while the phosphor screen is used for measurements of beam cross section. The H_2^+ beam is injected by means of a fast electrostatic

doi:10.1088/978-0-7503-4891-1ch2

Table 2.1. Nominal parameters of SIR [1].

Energy, $\boldsymbol{E}$	20 keV
H_2^+ Beam peak current, $\boldsymbol{I_0}$	5–25 μA
Circumference	6.57 m
No. of turns	< 200
Bunch length	0.15–5.5 m
Revolution period, $\boldsymbol{T}$	5 μs
Hor., vert. bare betatron tunes, $\boldsymbol{\nu_{x0}, \nu_{y0}}$	1.14, 1.11
Incoherent SC tune shift at 20 μA, $\boldsymbol{\delta\nu_{x,y}}$	0.005
Momentum compaction, $\boldsymbol{\alpha}$	1.0

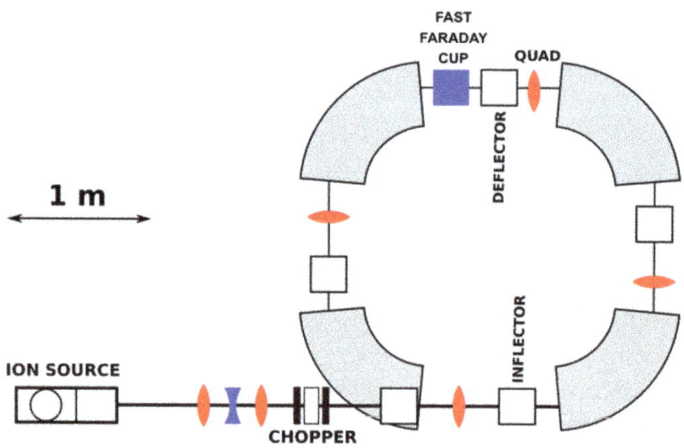

Figure 2.1. Top view schematics of the Small Isochronous Ring (SIR). Adapted with permission from [2].

Figure 2.2. Photograph of the Small Isochronous Ring at the Michigan State University laboratory in Lansing, MI. Reproduced from [1]. CC BY 3.0.

inflector that achieves isochronism by proper adjustment of gradient correctors in the bend magnets. Because of the isochronism, there is no de-bunching from longitudinal SC, but the energy spread increases in the 'frozen' longitudinal distribution. The number of turns is limited by charge neutralization from electrons in the residual gas. Furthermore, and unlike UMER (see chapter 1), the ambient field plays a minor role from the fact that the magnetic rigidity (see below) of the SIR beam is some 60 times that of the UMER beam (with both machines at the same energy, e.g., 20 keV).

2.2 Momentum compaction, slip factor, transition gamma, and chromaticity

We define the *momentum compaction* α_c as the fractional change in equilibrium radius per fractional change in linear momentum:

$$\alpha_c \equiv \frac{d\rho/\rho_0}{dp/p_0} = \frac{d\rho}{\rho_0}\delta^{-1}, \tag{2.1}$$

with p_0, ρ_0 defining design values of momentum and equilibrium radius, respectively. From the definition of *magnetic rigidity* as the ratio of momentum to particle's charge, $(B\rho) = p/q$, we have

$$\rho(p,\, B) = \frac{p}{qB},\ \rho_0 = \frac{p_0}{qB_0}, \tag{2.2}$$

and taking logarithmic derivatives on both sides we get $d\rho/\rho_0 = dp/p_0 - dB/B_0$. (Note that we are evaluating the derivatives, i.e., $\partial\rho/\partial p,\ \partial\rho/\partial B$, at the equilibrium values p_0, ρ_0.) Therefore, by using the definition of the *field index*, $n \equiv -(\rho_0/B_0)(\partial B_z/\partial r)_{r=\rho_0}$, we find:

$$\frac{d\rho}{\rho_0}(1 - n) = \frac{dp}{p_0}. \tag{2.3}$$

The momentum compaction is then

$$\alpha_c = \frac{1}{1 - n}. \tag{2.4}$$

Furthermore, because of the change in ρ there are corresponding changes in the revolution period $T = 2\pi\rho/\beta c$ and angular frequency $\omega = \beta c/\rho$. Taking logarithmic derivatives in the equation for ω, we get $d\omega/\omega_0 = d\beta/\beta_0 - d\rho/\rho_0$. The second term is $d\rho/\rho_0 = \alpha_c dp/p_0$, from equation (2.3), while the first term is $d\beta/\beta_0 = \gamma_0^{-2}dp/p_0$, which follows from the definition of the relativistic mass factor γ, and $d\gamma/\gamma_0 = \beta_0\gamma_0^2 d\beta$ [3]. Finally, the fractional change in period is

$$\frac{dT}{T_0} = -\frac{d\omega}{\omega_0} = \eta_{tr}\frac{dp}{p_0},\ \text{with}\ \eta_{tr} = \alpha_c - \frac{1}{\gamma_0^2} = \frac{1}{\gamma_{tr}^2} - \frac{1}{\gamma_0^2}. \tag{2.5}$$

The last equality defines the *slip factor* η_{tr}, and the *transition gamma*, γ_t,

$$\gamma_t = \frac{1}{\sqrt{\alpha_c}}. \tag{2.6}$$

In a straight machine, $\alpha_c = 0$, $\gamma_t \to \infty$, $\eta_{tr} = -1/\gamma_0^2$: at all energies, particles with higher momentum take less time to traverse a given distance. In circular machines, the same would apply for energies such that $\gamma_0 < \gamma_t$. However, there are situations where, counterintuitively, particles with higher momentum take *more* time to complete a revolution, i.e., $\eta_{tr} > 0$, or $\gamma_0 > \gamma_t$. This is the regime of 'negative mass' of weak focusing machines such as cyclotrons, or strong focusing circular machines above transition energy. For cyclotrons we have $0 < n < 1$, $\nu_r < 1$ (see [4], chapter 2), while $\nu_r > 1$ for strong focusing machines like UMER or SIR.

Ideally, in the isochronous regime, the slip factor η_{tr} defined in equation (2.5) is zero indicating that the revolution period is independent of energy. Furthermore, at low energy, $\gamma_0 \approx 1$ leading to a momentum compaction factor $\alpha_c \approx 1$. The actual slip factor in SIR is 2×10^{-4} [1].

In circular machines, another effect of a fractional momentum error $\delta = \delta p/p_0$ is to change the tunes in both the radial and vertical planes. Horizontal and vertical *chromaticities*, ξ_X, ξ_Y, are defined to characterize the effect:

$$\Delta\nu_X = \xi_X\delta, \quad \Delta\nu_Y = \xi_Y\delta. \tag{2.7}$$

The focusing of particles with higher momentum is less efficient, leading to reduced tunes. Therefore, we expect the chromaticity from linear elements such as dipoles and quadrupoles to be negative.

In a weak focusing machine where $\partial B_y/\partial\rho$ is constant, the field index satisfies $dn/d\rho = n(1 + n)/\rho$ (see [5]) and we find

$$\xi_X = -\frac{n(n+1)}{2(1-n)^{3/2}}, \quad \xi_Y = -\frac{\sqrt{n}(n+1)}{2(1-n)}. \tag{2.8}$$

A more general definition of chromaticity, referred to as the *natural chromaticity* is [4]

$$\xi_X = -\frac{1}{4\pi}\oint \beta_X(s)\kappa_X(s)ds, \quad \xi_Y = \frac{1}{4\pi}\oint \beta_Y(s)\kappa_Y(s)ds, \tag{2.9}$$

where only linear focusing elements are involved. These expressions are easily derived by calculating the first-order differential change in tune when the one-turn matrix is multiplied by a differential-kick error matrix; this latter kick arises from a small fractional momentum error δ. Details can be found in e.g., [6].

2.3 Linear dispersion and direct space charge (SC)

The single-particle equations of motion are modified in the presence of a momentum spread Δp around the design value p_0. For the horizontal motion (radial in a circular machine), we have the inhomogeneous equation (see, e.g., [7] p 46)

$$x''(s) + k_{0x}^2(s)x = \frac{1}{\rho(s)}\delta, \tag{2.10}$$

where $\delta = \Delta p/p_0$, and $\rho(s)$ is the *local* orbit radius. A major assumption here is that $x(s) \ll \rho(s)$, so we can neglect the *weak focusing* term $1/\rho^2$. The general solution of equation (2.10) is

$$x(s) = x_\beta(s) + D_{0x}(s)\delta, \tag{2.11}$$

where $x_\beta(s)$ is the solution of the *homogeneous* version of equation (2.10), and $D_{0x}(s)$ defines the *linear dispersion function.* $D_{0x}(s)$ satisfies the equation (see, e.g., [7, p 49])

$$D_{0x}''(s) + \left[\frac{1}{\rho^2(s)} + k_{0x}^2(s)\right]D_{0x}(s) = \frac{1}{\rho(s)}. \tag{2.12}$$

Normally, the weak focusing term $1/\rho^2(s)$ can be neglected as in equation (2.10). Thus, using the smooth-approximation whereby $D''_{0x}(s) \to 0$, we obtain for the *average dispersion*:

$$\overline{D}_{0x} = \frac{1}{k_{0x}^2\rho_m} = \frac{\rho_m}{\nu_{0x}^2}, \tag{2.13}$$

with the last equality obtained from equation (1.6). Therefore, large circular machines normally need to operate at large tunes to reduce dispersion effects. However, given SIR compact size, it can operate with small tune values (see table 2.1).

With direct SC, we could write as an ansatz, $\overline{D}_x = \rho_m/\nu^2 = \overline{D}_{0x}/\eta^2$ where η is the tune depression, equation (1.9) (not to be confused with the slip factor η_{tr} of equation (2.5)); thus, the average linear dispersion should increase because of space charge. However, the last expression overestimates the effect, but a quantity related to η should be used. We find (see [4], section 7.5)

$$\overline{D}_x = \frac{\overline{D}_{0x}}{\eta_{x\Delta}^2}, \quad \eta_{x\Delta} \equiv \sqrt{1 - \frac{K}{a^2k_{x0}^2}}. \tag{2.14}$$

Furthermore, for weak SC we have $K/a^2k_{0x}^2 \ll 1$ so

$$\frac{\delta D}{\overline{D}_{0x}} \cong \frac{K}{a^2k_{0x}^2} = 2\frac{\delta\nu}{\nu_{0x}}, \tag{2.15}$$

where $\delta\nu/\nu_{0x}$ is the fractional incoherent SC tune shift, i.e., left-hand side in equation (1.8), divided by the bare betatron tune.

We can now relate dispersion to the momentum compaction factor. If C denotes the length of the orbit for a particle of coordinate x, and C_0 is the length of the reference orbit (for which $x = 0$ by definition), then the momentum compaction, equation (2.1), is $\alpha_c = (dR/\rho_0)\delta^{-1} = (dC/C_0)\delta^{-1}$. Therefore, we can write, to first order in x (see, e.g., [8, 9])

$$\alpha_c = \frac{1}{C_0}\int_0^{C_0}\frac{D_{0x}(s)}{\rho(s)}ds, \tag{2.16}$$

since $dC = C - C_0 = \int_0^{C_0}(x/\rho)ds$, [$\rho(s)$ = local radius of curvature], and $x(s) = D_{0x}(s)\delta$, from equation (2.11) with $x_\beta = 0$.

Research in SIR has revealed a fast-growing SC driven instability as illustrated by the measured longitudinal bunch profiles shown in figure 2.3. The instability develops over a time scale that is slow compared to betatron oscillations. It can be explained by transverse-longitudinal coupling whereby deformation of the beam shape ('snaking') leads to a coherent SC field that affects the coherent longitudinal motion. In effect, the deformation of the beam shape enhances a *negative-mass-like instability*—see next section. The latter occurs if the slip factor is positive, i.e., above the transition energy. However, SC changes the dispersion function, as we just discussed, which in turn changes the slip factor. From equation (2.16) for the momentum compaction, we can write

$$\alpha_{cSC} = \frac{1}{C_0}\int_0^{C_0}\frac{(D_0 + \delta D)}{\rho(s)}ds = \alpha_c + \frac{\delta D}{R_0}, \tag{2.17}$$

where we have included the change in dispersion function from incoherent, i.e., direct SC, and R_0 is the average ring radius $R_0 \equiv \rho_m$. Therefore, from equations (2.5) and (2.17) we have (see also [1])

$$\delta\eta = \delta\alpha_{cSC} = \frac{\delta D}{R_0}. \tag{2.18}$$

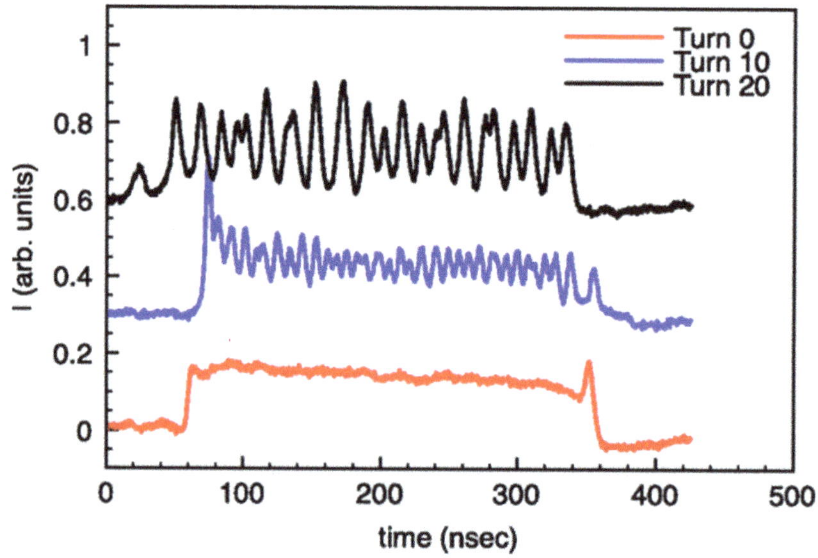

Figure 2.3. Longitudinal bunch profiles in SIR. The beam energy and peak current are 20.9 keV and 9.3 μA. Reproduced from [1]. CC BY 3.0.

Pozdeyev *et al* [1] model the 'snaking' of the beam by assuming a small distortion and no image forces. Then, separate equations for the betatron and centroid motions in a uniform-focusing approximation of the ring lead to a steady state solution for the centroid part. The net result from the latter is that the change in average dispersion (from its value R_0/ν_0^2) is $\delta D > 0$ as in equations (2.15) and (2.18). Therefore, the instability can happen even if the ring is operating below transition.

Regarding $\delta\nu/\nu_{0x,}$ the expression quoted in equation (9) of [1] can be re-written in SI units as

$$\frac{\delta\nu}{\nu_0} \cong \frac{1}{4\pi\varepsilon_0}\frac{q\Lambda_0}{\gamma^3 m\omega^2 a^2}\frac{1}{\nu_0^2}, \tag{2.19}$$

where Λ_0 is the line charge density, a is the average beam radius, and $\omega^2 = \beta^2 c^2/R_0^2$; we have used results from appendix A of [2]. Equation (2.19) can be easily shown to be equivalent to our equation (1.8) (see also equation (1.2)), since $a^2 = \tilde{\varepsilon}_x\overline{\beta} = \tilde{\varepsilon}_x/k_0$, $\nu_0 = R_0 k_0$, and $I = \Lambda_0\beta c$ ($\overline{\beta}$ is the average betatron function, not to be confused with the relativistic parameter β.)

2.4 Negative-mass instability

Ideally, a coasting beam in an accelerator would be propagated at the speed of light, and inside a smooth, perfectly conducting cylindrical vacuum pipe. At the light speed, the beam self- electric and magnetic fields would completely balance one another. In reality, the cancellation is not perfect at sub-light speeds and far from it at low energies. In addition, the beam encounters many unavoidable discontinuities in a vacuum pipe and the latter, by necessity, has a finite conductivity. Thus, the interaction of the beam with its environment leads to the appearance of wakefields, i.e., radiation fields of the image charges that can interact back with the beam and cause damaging instabilities, especially in circular accelerators. The theoretical general treatment of wakefield-beam interactions involves the solution of the coupled Vlasov–Maxwell equations [5, 10], but a simpler approach is to introduce concepts borrowed from circuit theory such as impedance. This approach is also useful for the study of the *negative-mass instability* (NMI). Above transition energy in a synchrotron, particles in front and near a longitudinal charge perturbation (a 'bump') are accelerated by longitudinal SC forces, but take longer to complete a revolution than reference particles. Particles behind the bump, on the other hand, are decelerated but take less time. Thus, in the absence of other factors, the original perturbation can grow indefinitely leading to an instability; this is the essence of the NMI.

In this section, we follow Wiedemann [11] to derive the growth rate of NMI, which is also treated by Reiser [5] in his section 6.3.3 in a more general context. Additional useful material can be found in Chao's 2017 CERN School notes [12], his classic monograph [10], and the articles by Zotter [13, 14].

Let us consider a perturbation in the longitudinal line charge density $\Lambda(s,t)$ of a coasting beam in a circular machine of average radius R_0. If $\lambda = 2\pi/k$ is the wavelength of the perturbation, we assume that there are n such wavelengths in one revolution, i.e., we focus our attention on the nth harmonic. Therefore, if Ω is the frequency of the perturbation, $ks - \Omega t = n\theta - \Omega t$ since $n\lambda = 2\pi R_0$, or $n = kR_0 = ks/\theta$. With this understanding, we can write the following equations for the line charge density, revolution frequency and beam current:

$$\Lambda(\theta,\, t) = \Lambda_0 + \Lambda_n e^{i(n\theta-\Omega t)}, \tag{2.20}$$

$$\omega(\theta,\, t) = \omega_0 + \omega_n e^{i(n\theta-\Omega t)}, \tag{2.21}$$

$$I(\theta,\, t) = I_0 + I_n e^{i(n\theta-\Omega t)}. \tag{2.22}$$

The symbols Λ_0, ω_0, I_0 represent the unperturbed (DC) quantities, and Λ_n, ω_n, I_n the amplitudes characterizing the perturbation's nth harmonic. The rate of change of the revolution frequency ω can be related to the slip factor η_{tr} (equation (2.5)) and the rate of change of energy E, for particles near the perturbed line distribution:

$$\frac{d\omega}{dt} = -\frac{\eta_{tr}\omega_0}{\beta_0^2 E_0}\frac{dE}{dt}, \tag{2.23}$$

which can be derived from expressions for $d\gamma/\gamma_0$ and $d\beta$ quoted in connection to equation (2.5). (Notice that Wiedemann defines η_{tr} such that it is the negative of our own—see his equation (5.132) in [11] and compare it to our equation (2.5). Furthermore, he calls $-\eta_{tr}$ 'momentum compaction', unlike our usage.) Below transition, i.e., $\eta_{tr} < 0$, the revolution frequency increases with increasing energy E. The rate of change of E, on the other hand, can be expressed in terms of the voltage drop (from the perturbation nth harmonic) in one revolution $V_n = -2\pi R_0 E_{\|} = Z_{\|} I_n$, with $E_{\|}$, $Z_{\|}$ denoting the longitudinal electric field and longitudinal *space charge impedance*, respectively. Thus, we have

$$\frac{dE}{dt} = -qV_n\frac{\omega_0}{2\pi} = -qZ_{\|}I_n e^{i(n\theta-\Omega t)}\frac{\omega_0}{2\pi}. \tag{2.24}$$

Therefore, from equations (2.21) and (2.23) we obtain

$$\omega_n(\Omega - n\omega_0) = i\eta_{tr}\frac{q\omega_0^2}{2\pi}\frac{I_n Z_{\|}}{\beta_0^2 E_0}. \tag{2.25}$$

The 1D continuity equation in the form $\partial\Lambda/\partial t + \partial(\beta c\Lambda)/\partial z = 0$, or, since $\beta c/R_0 = \omega$,

$$\frac{\partial\Lambda}{\partial t} + \omega\frac{\partial\Lambda}{\partial\theta} + \Lambda\frac{\partial\omega}{\partial\theta} \cong \frac{\partial\Lambda}{\partial t} + \omega_0\frac{\partial\Lambda}{\partial\theta} + \Lambda_0\frac{\partial\omega}{\partial\theta} = 0, \tag{2.26}$$

can be used now to obtain $(\Omega - n\omega_0)\Lambda_n = n\omega_n\Lambda_0$, from equations (2.20) and (2.21). Finally, we use the expression for ω_n from the last equality, and equation (2.25), to write

$$(\Omega - n\omega_0)^2 = i\eta_{tr}\frac{nq\omega_0^2}{2\pi}\frac{I_0 Z_{\|}}{\beta_0^2 E_0}. \tag{2.27}$$

If $\Delta\Omega = \Omega - n\omega_0 = \mathrm{Re}(\Delta\Omega) + i\mathrm{Im}(\Delta\Omega)$, $\eta_{tr} > 0$, and $Z_{\|}$ is positive imaginary [11], i.e., 'capacitive' as for the SC impedance, the growth rate of the *negative-mass instability* is given by

$$\tau^{-1}(k) = \mathrm{Im}(\Delta\Omega) = \omega_0\sqrt{\eta_{tr}\frac{qI_0kR_0|Z_{\|}|}{2\pi\beta_0^2E_0}}, \tag{2.28}$$

since $n = kR_0 = 2\pi R_0/\lambda$. Because the slip factor η_{tr} is increased by the larger dispersion from direct SC, equation (2.18), the instability can happen even if the optics are set up for below transition. The instability is a *microwave instability* in the short wavelength limit of the longitudinal *SC impedance* $Z_{\|}$ (see [14] for definition). At the transition energy, i.e., with isochronous operation, the rate is simply proportional to the peak beam current [1].

2.5 Simulations and experiments

Simulations were conducted using the code CYCO [2] that included a particle-in-cell (PIC) solver to calculate the SC fields, with initial effective phase-space beam distributions based on experimental results (uniform 40 cm long bunch with hard edges, and initial zero energy spread). Two regimes were studied, isochronous and below transition. It was found that a harmonic of the linear charge oscillations, for 10 μA peak current, with a wavelength of the order of the beam diameter (1.5 cm) displays the fastest growth rate of the instability. Further, a simple theoretical model derived from equation (2.28) in the isochronous regime, after correcting for the 'SC depressed' slip factor, yields growth rates that are some 20% higher than results from simulations; both theory and simulations, however, predict rates that are much higher than results based on a model for the microwave instability alone. As an example, at 10 μA peak current, the growth rate is 0.29 turn^{-1} from simulations, 0.35 turn^{-1} from the adjusted equation (2.28), and only 0.028 turn^{-1} using the pure microwave instability formula, equation (2.28). Figure 2.4 shows results of simulations in the isochronous regime for 10 μA peak current.

Lastly, the simulations agree very well with experiments. The latter were conducted in the near-isochronous regime (slip factor $\eta_0 = 2 \times 10^{-4}$) and were based on counting the number of peaks of the longitudinal charge density as a

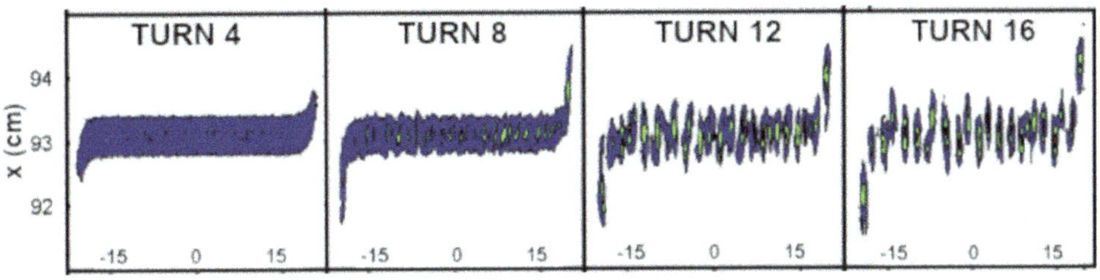

Figure 2.4. Simulations with CYCO of the evolution of a 10 μA, 40 cm bunch in the isochronous regime: projections of the bunch charge density on the median plane. '*x*' (vertical axis) is measured from the center of the ring. Adapted from [1]. CC BY 3.0.

function of turn number. From theoretical grounds, the growth rate of the instability is linearly proportional to the beam current, with no dependence on bunch length in the experiments.

2.6 Prospects

We quote Pozdeyev *et al* from a PAC2005 publication [15]: 'Because the ring is isochronous, it does not require nor include an RF system. The simplicity of the lattice design allowed us to build the ring in a short time with limited resources without compromising the goal of experimental studies of space charge effects in the isochronous regime.'

Upgrade objectives were listed back then in 2005: increase the SC intensity to study resonance traversal, beam lifetime, head-tail and beam envelope instabilities. For this, the beam peak current would go up to 230 μA (still at 30 keV maximum energy); the circumference would be increased from 6.6 m to 8.6 m by extending the four straight sections by 0.5 m (figure 2.1). Further, two new quadrupoles would be added per straight section, a simple RF system implemented for longitudinal focusing, a high-resolution energy analyzer (similar to one designed for the UMER project) installed, and two sets of sextupole magnets included for chromaticity corrections. The new regime of research in SIR would be non-isochronous, and the fractional Laslett tune shift (fractional incoherent SC tune shift—see chapter 1) would go from a few percent to 50%.

Now that SIR resides at the U. of Maryland, but not directly connected to the UMER group, we will be looking forward to the implementation of these or similar upgrades.

2.7 Computer resources

The MATHCAD worksheet ***SIR-BasicCalc.xmcd*** (see author's website[1]) contains calculations of incoherent SC tune shift based on data from table 2.1 and Pozdeyev's PhD thesis [2].

References

[1] Pozdeyev E, Rodriguez J A, Marti F and York R C 2009 Longitudinal beam dynamics studies with space charge in small isochronous ring *Phys. Rev. ST. Accel. Beams* **12** 054202

[2] Pozdeyev E 2003 CYCO and SIR: new tools for numerical and experimental studies of space charge effects in the isochronous regime *Dissertation* Department of Physics and Astronomy, Michigan State University https://groups.nscl.msu.edu/nscl_library/Thesis/Pozdeyev,%20Eduard.pdf

[3] Bovet C, Gouiran R, Gumowski I and Reich K H 1970 A selection of formulae and data useful for the design of A.G. synchrotrons *CERN Technical Note DL/70/4*

[4] Bernal S 2022 *A Practical Introduction to Beam Physics and Particle Accelerators* 3rd edn (Bristol: IOP Publishing)

[1] https://ireap.umd.edu/clark/faculty/1273/Santiago-Bernal

[5] Martin R 2008 *Theory and Design of Charged Particle Beams* 2nd edn (Weinheim: Wiley-VCH)
[6] Guiducci S *Chromaticity* (CAS CERN Accelerator School) CERN 91-04 pp 53–66
[7] Bryant P J and Johnsen K 1993 *The Principles of Circular Accelerators and Storage Rings* (Cambridge: Cambridge University Press)
[8] Wolski A 2014 *Beam Dynamics in high Energy Particle Accelerators* (London: Imperial College Press)
[9] Wille K 2000 *The Physics of Particle Accelerators, An Introduction* (New York: Oxford University Press)
[10] Wu Chao A 1993 *Physics of Collective Beam Instabilities in High Energy Accelerators* (New York: Wiley)
[11] Wiedemann H 2007 *Particle Accelerator Physics* 3rd edn (Berlin: Springer)
[12] Chao A 2017 Beam dynamics of collective instabilities in high-energy accelerators *CERN Yellow Reports: School Proc.* CERN-2017-006-SP (Geneva: CERN)
[13] Zotter B 2013 Single-bunch instabilities in circular accelerators *Handbook of Accelerator Physics and Engineering* ed A W Chao, K H Mess, M Tigner and F Zimmermann (Singapore: World Scientific) 2nd edn
[14] Zotter B 2013 Space charge effects in circular accelerators *Handbook of Accelerator Physics and Engineering* ed A W Chao, K H Mess, M Tigner and F Zimmermann (Singapore: World Scientific) 2nd edn
[15] Pozdeyev E, Marti F, York R C and Rodriguez J 2005 Future plans for the small isochronous ring *IEEE Proc. 2005 Particle Accelerator Conf. (Knoxville, TN)* 1778

Chapter 3

Nonlinear optics and other physics at the Integrable Optics Test Accelerator (IOTA)

The IOTA proton/electron ring is currently in development and operation at Fermilab National Laboratory; it is part of Fermilab's Accelerator Science and Technology Facility (FAST). The mission of IOTA is the investigation of novel nonlinear optics lattices, optical stochastic cooling, space charge compensation, and other techniques, all with relevance to the intensity frontier of accelerators for nuclear and particle physics.

We describe in section 3.1 IOTA's physical configuration and parameters. In section 3.2, we summarize the ideas behind nonlinear integrable and quasi-integrable optics (NIO), considered the main thrust of IOTA, as it involves novel nonlinear optics techniques to circumvent resonances and other factors that affect beam quality in electron and hadron rings. We also discuss briefly in this section the electron lens, which complements the studies of NIO with octupole lenses.

We begin section 3.3 with a short description of beam cooling techniques in general, and then describe in some detail standard stochastic cooling and the recently demonstrated optical stochastic cooling (OSC) in IOTA.

In section 3.4, we summarize a recent publication by the IOTA team and collaborators on the use of synchrotron radiation (SR) noise to measure transverse emittances of round and flat electron beams.

In section 3.5, we describe space charge compensation (SCC), other experiments in IOTA and additional work with a single electron. In the last two sections, we cover IOTA prospects and computer resources.

3.1 IOTA lattice, parameters and diagnostics

The IOTA ring and its components, shown schematically in figure 3.1, are covered in detail in several publications in the *Journal of Instrumentation*. Antipov *et al* [1] review in detail the physics program in IOTA, and the ideas behind quasi-integrable

doi:10.1088/978-0-7503-4891-1ch3

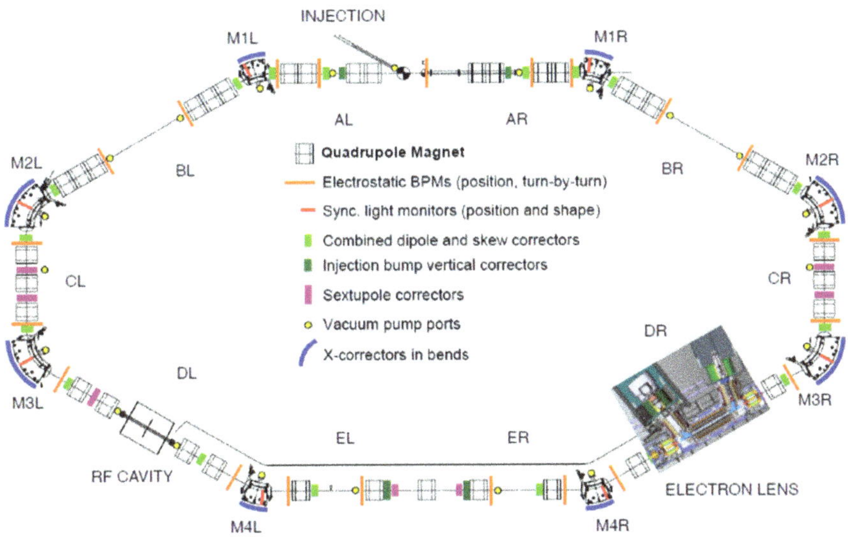

Figure 3.1. Schematic layout of the IOTA ring. The straight sections are labeled AR, BR, etc (A Right, B Right, etc) and AL, BL, etc (A Left, B Left, etc). Reproduced from [3] with minor changes, copyright IOP Publishing Ltd. CC BY 4.0.

nonlinear optics [2]. Stancari *et al* [3] discuss the electron lens, a major component of IOTA. Valishev *et al* [4] cover the first results of experiments, while Romanov *et al* [5] report on experiments with a single electron. Most of the material that follows in this section is a summary of the latter references.

As shown in figure 3.1, there are eight main bending dipole magnets, 39 quadrupole magnets, 10 sextupoles and 20 dipole and skew-quadrupole correctors. The 40 m circumference, 50 mm aperture ring is designed for operation with either electron or proton beams. The 150 MeV electron beam is used for the integrable optics tests, which require demanding control of the focusing lattice and orbits, with at least 1% accuracy of β-functions. This control is feasible with moderately relativistic electrons ($\gamma = 294$) for which SC effects are negligible. For 2.5 MeV protons, on the other hand, SC effects are significant at the IOTA operating parameters. Table 3.1 summarizes the main parameters for both electron and proton beam operations in IOTA.

For electron operations, the IOTA ring is preceded by a 125 m linac consisting of a 5 MeV RF photoinjector, a 50 MeV low-energy beamline, and a 100 m, approximately, 300 MeV section. The electron gun is equipped with two solenoids for *emittance compensation* [7] or for the production of angular-momentum dominated beams (see e.g., [8]). The low-energy section consists of two superconducting RF cavities, focusing and steering elements, an optional chicane, and a spectrometer dipole that steers the beam into a beam absorber, or lets the beam into the high-energy section when unpowered. The high-energy beamline has eight superconducting cavities that can provide acceleration to 250 MeV. It also contains a FODO lattice and a dogleg to direct the beam into test sections and absorbers, or

Table 3.1. Nominal parameters of IOTA [1, 3].

Parameter	Electrons	Protons
Max. kinetic energy, $\boldsymbol{E}$	150 MeV	2.5 MeV
Circumference, $\boldsymbol{C}$	39.97 m	39.97 m
Revolution period, $\boldsymbol{T}$	133.3 ns	1.83 μs
Number of particles, peak beam current	2×10^9, 2.4 mA	9×10^{10}, 8 mA
Bunch length	10.8 m	1.7 m
Equilibrium norm. emittance, ε_x, ε_y	0.04, 0.04 μm	0.3, 0.3 μm
RF voltage, frequency, harmonic no.	1 kV, 30 MHz, 4	400 V, 2.18 MHz, 4
Hor., vert. bare betatron tunes, $\boldsymbol{\nu_{x0}, \nu_{y0}}$	4–6	6
Max. incoh. SC tune shift (unbunch., bunched)	$< 10^{-3}$	0.5, 1.2

into the IOTA ring. For proton operation, IOTA's injector is based on a pulsed (1 Hz maximum), 2.5 MeV radio-frequency quadrupole (RFQ). Because of the strong longitudinal SC of the low-energy proton beam, without RF complete de-bunching would occur in less than one turn (see chapter 6 in [8] for a de-bunching calculation in UMER). At a pressure of 6×10^{-10} Torr, the lifetime of the proton beam is expected to be 5 min; at 3×10^{-8} Torr, the electron beam lifetime should be 30 min. These lifetimes are calculated at the energies shown in table 3.1.

Diagnostics include capacitive beam position monitors (BPMs), originally developed for EMMA (chapter 4), in the ring (20) and electron injector, with better than 50 μm resolution for a single bunch. In addition, there is a Faraday cup downstream of the RF injector, and wall-current monitors (WCMs) and toroids for absolute calibration of the WCMs. Two WCMs and two toroids are placed upstream of the high-energy section, and two toroids downstream. An additional WCM is planned for the ring. There are also ten YAG:Ce-crystal screens for transverse beam profiling in the electron injector. The screens are used in conjunction with multi-slit masks for transverse emittance measurements. Finally, longitudinal beam diagnostics of the electron beam utilize a streak camera to analyze optical and THz radiation from synchrotron and transition radiation.

Special diagnostics for the proton beam are also present, including ionization profile monitors. It is to be noted that challenges remain regarding the use of the same RF system and diagnostics (e.g., BPMs) for both electrons and protons. IOTA started operation with electrons in August 2018 and was undergoing the installation of the proton injector at the end of 2021.

3.2 Nonlinear Integrable and quasi-integrable optics (NIO)

The betatron tunes in a circular lattice of (nominally) linear elements such as quadrupoles are independent of betatron amplitude. Beams in such a lattice are subject to linear and nonlinear resonances (see chapter 7 in [8]), the latter from nonlinear field components in the actual magnets. In addition, the presence of nonlinear elements such as sextupoles and/or octupoles (see appendix B in [8]) creates amplitude-dependent tunes that may help beam stability but also tend to reduce the dynamic aperture (DA). It is possible, however, to construct special lattice inserts with nonlinear fields such that the *single-particle* Hamiltonian leads to integrable or quasi-integrable (QI) conditions. In the former case, two integrals of the motion are present: the total energy and a second one, a nonlinear analog of the C–S invariants. For QI conditions, only the total energy is invariant, but the resulting tune spread from the nonlinear fields is expected to enhance beam stability through the mechanism of *Landau Damping* [9]. An example of a QI octupole insert in UMER was described in chapter 1; IOTA has also implemented QI inserts either with a quasi-continuous octupole channel, or with a small number of separated octupole magnets [10]. Furthermore, a fully integrable system based on a 'quasi-continuous nonlinear elliptic-potential focusing channel' is being tested in IOTA. This insert follows the ideas of Danilov–Nagaitsev (DN) [11].

As an example of a nonlinear QI potential, we consider an octupole channel. The nonlinear part of the single-particle 2D Hamiltonian, $V(x,y,s)$, can be written as [11]

$$V(x, y, s) = \frac{\kappa}{4\beta(s)^3}(x^4 + y^4 - 6x^2y^2), \tag{3.1}$$

where $\beta(s) \equiv \beta_X(s) = \beta_Y(s)$ is the betatron function and κ is the octupole strength. The dependence on β assures that the potential in equation (3.1) is time-independent in *normalized phase space variables*, x_N, y_N, ψ. We write $U(x_N, y_N, \psi) = \beta(\psi)V[x_N\sqrt{\beta(\psi)}, y_N\sqrt{\beta(\psi)}, s(\psi)]$, so

$$U(x_N, y_N) = \frac{\kappa}{4}\left(x_N^4 + y_N^4 - 6x_N^2y_N^2\right). \tag{3.2}$$

The second invariant in one of the DN integrable potentials is a quadratic function of momenta. It involves arbitrary functions (f_2, g_2) of elliptic variables (ξ, η) that

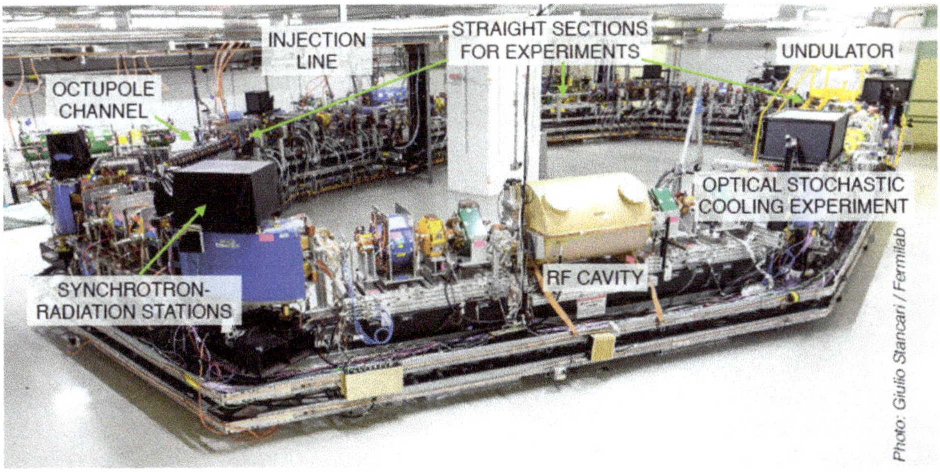

Figure 3.2. The IOTA ring in May 2021. Photo: Giulio Stancari/Fermilab [6], with permission.

depend on an arbitrary constant c. *Omitting the subscript N*, the potential can be written as

$$U(x, y) = \frac{x^2 + y^2}{2} + \frac{f_2(\xi) + g_2(\eta)}{\xi^2 - \eta^2}, \tag{3.3}$$

where $f_2(\xi) = \xi\sqrt{\xi^2 - 1}(d + t \text{ acosh}(\xi))$, $g_2(\eta) = \eta\sqrt{1 - \eta^2}(q + t \text{ acos}(\eta))$,with d, q, and t representing arbitrary constants. The first few terms of the potential for $c = 1$ are [12]

$$U(x, y) = \frac{1}{2}(x^2 + y^2) + t\text{Re}\left[(x + iy)^2 + \frac{2}{3}(x + iy)^4 + \frac{8}{15}(x + iy)^6 + \frac{16}{35}(x + iy)^8 + \cdots\right], \tag{3.4}$$

where t is the magnitude of the nonlinear part of the potential. Stable small-amplitude motion requires $0 \leqslant t < 0.5$. The corresponding betatron tunes in normalized coordinates are $\nu_{X,\,Y} = \nu_0\sqrt{1 \pm 2t}$. For $t \to 0$, the tunes are $\nu_X \to \nu_Y \to \nu_0$, the linear unperturbed tune, while $\nu_X \to \sqrt{2}\,\nu_0$, $\nu_Y \to 0$ for $t \to 0.5$. Thus, tune spreads of 100%, 40% in the vertical, horizontal planes, respectively, would be possible in principle. The potential in equation (3.4) is shown in figure 3.3.

Tests of the NIO began in 2019 and continued through 2020 ('run 2'). The QI experiments rely on a 1.8 m insert with 17 discreet octupoles to approximate a continuous potential. A tune spread of 0.04 was achieved, with observed detuning in agreement with tracking simulations [4]. The threshold of instabilities versus

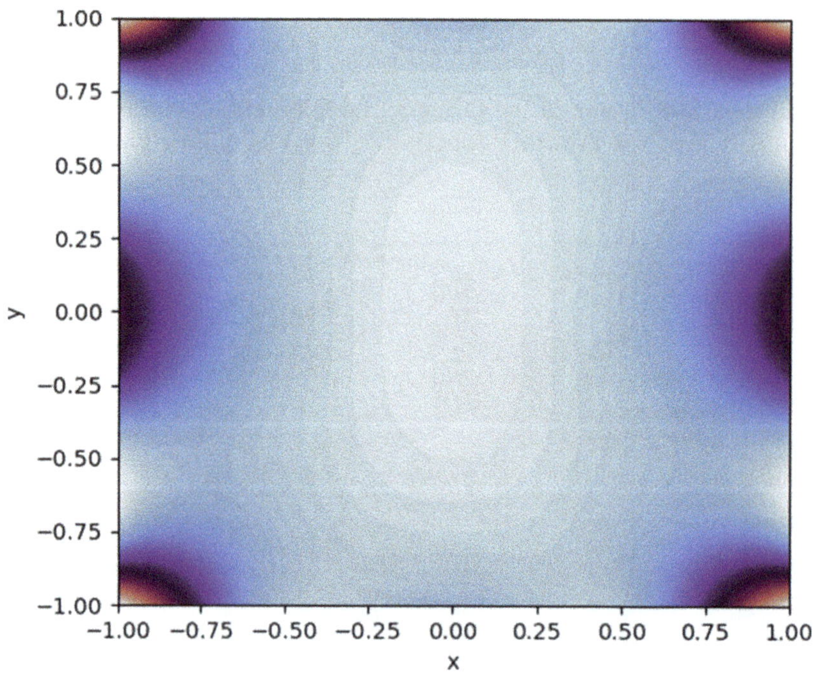

Figure 3.3. Contour plot of the potential in equation (3.4) with $(c, t) = (1.0, 0.4)$. Singularities occur at $(x, y) = (\pm 1.0,\ 0)$.

strength of nonlinearity was also measured with the QI insert to test for Landau damping. There was a doubling of the instability threshold with octupole current, clearly showing the stabilizing effect of the tune spread. The runs with the DN configuration, on the other hand, use a specially designed 1.8 m insert with 18 independent magnets with varying apertures. The effect of nonlinearities on betatron resonances as a function of the t parameter was studied; the 100 MeV electron beam could circulate for 110 s at $t = 0.5$, i.e., on the integer resonance. An effect of crossing the integer resonance is the splitting of the beam into two stable beamlets in the vertical plane, in agreement with theory.

Experiments with a nonlinear electron lens are also planned to investigate further integrable optics with both circulating electron and proton beams. The electron lens will be installed in section DR of IOTA (figure 3.1); the schematics are shown in figure 3.4.

The electron lens consists of a low-energy (100 eV–10 keV), high-current electron gun, beam collector, solenoids at the gun and collector, and a long solenoid (0.7 m) on the main IOTA beamline. Two ideas for NIO are under consideration: (1) a McMillan lens where the electron beam has a specific current distribution that interacts with the circulating beam to produce 'thin' nonlinear radial kicks that depend on betatron amplitude; (2) an axially symmetric, thick electron lens that will operate in a region of equal and constant betatron functions in the ring. The second option is more robust than the McMillan lens as it works with any electron lens current distribution, although the resulting tune spread is smaller than in the first option. The tune spread if the thick lens is used can be easily estimated [1, 3] as

$$\Delta\nu = \frac{1}{(B\rho)}\frac{B_z L}{4\pi} = \frac{L}{2\pi\beta_{\text{lat}}}, \tag{3.5}$$

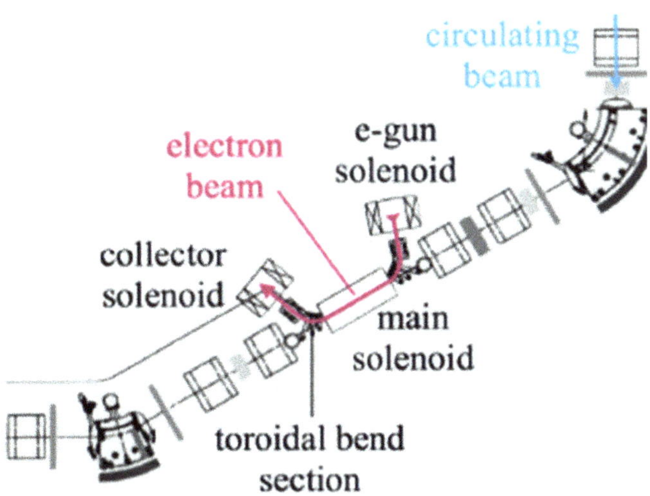

Figure 3.4. General layout of the IOTA electron lens. Reproduced from [1], copyright IOP Publishing Ltd. CC BY 3.0.

where $(B\rho)$ is the magnetic rigidity, B_z is the long solenoid B-field, L is the solenoid's length, and $\beta_{\text{lat}} \equiv \beta_X = \beta_Y$ is the lattice (constant) beta function in the long solenoid section. For 150 MeV electrons, $(B\rho) = 0.50$ T-m; since $L = 0.7$ m, and assuming $\beta_{\text{lat}} = 4$ m, $B_z = 0.50$ T, we get $\Delta\nu = 0.028$.

3.3 Beam cooling techniques and optical stochastic cooling (OSC)

Beam cooling consists in the manipulation of a beam's particle distribution in order to reduce emittance and thus increase *brightness*. The latter is, in the transverse plane, $\mathrm{B} \propto I_\mathrm{p}/(\varepsilon_x\varepsilon_y)$, i.e., proportional to the phase space density (I_p is the beam peak current.) A good overview of beam cooling is given in [13, 14]. Table 3.1 of [13] summarizes the main features of the different methods. The applicability of the latter depends on particle species (electrons versus ions), machine configurations, energies, and beam current and dimensions. A beam can be cooled with another co-moving beam (e.g., cooling of a heavy-ion beam with an electron beam); through processes such as ionization or laser cooling; by electronic-feedback methods (stochastic cooling); or by radiative cooling which involves damping motion via a combination of synchrotron radiation, random quantum photon emission and RF acceleration. Radiative or radiation cooling is used in all synchrotron light facilities and in damping rings for beam injection into electron linear colliders.

In *electron cooling*, suggested by Budker in 1966, heat flows from the hotter (ion) beam to the colder (electron) one in an irreversible way. At equilibrium, the two kinetic beam temperatures are equal, but the emittance of the ion beam is much smaller than that of the electron beam. The method cools all degrees of freedom simultaneously and works best for low energy anti-protons and ions; typical cooling times are milliseconds to hours.

In *ionization cooling*, introduced by Skrinsky around 1980, energy is deposited to a material medium, but longitudinal momentum is recovered in RF cavities as in radiation damping (see appendix A); this process is envisioned as a way to cool muon beams. Ionization cooling is limited by statistical fluctuations (straggling) and random multiple scattering.

In *laser cooling*, two counter-propagating laser beams are tuned to special closed energy transitions of certain ions; cooling occurs in the longitudinal plane, and it is counteracted by heating via photon spontaneous emission. Laser cooling achieves the lowest ion beam temperatures, and its main long-term goal is the production of crystalline beams. It is also applicable in the Paul traps described in chapter 5 of this book.

Common features of all cooling methods are the presence of external control agents (other beams, ionizing media, lasers, kicker-pickup feedback electronics, focusing/bending lattice, RF cavities, etc), and heating mechanisms (spontaneous emission, straggling, scattering, incoherent SC, intra-beam scattering, noise instrumentation, etc) leading to dissipation, i.e., a non-Liouvillean process. The balance between the two mechanisms yields the equilibrium emittance(s). The existing theories of beam cooling rely on kinetic models and statistical considerations and are, for the most part, successful at describing experimental situations.

From the point of view of thermodynamics and statistical mechanics, *stochastic cooling* is perhaps the most interesting method as it relies on a scheme reminiscent of the cooling of a gas by Maxwell's demon. In Maxwell's thought experiment, the demon actuates a gate in a hole at a partition between two vessels containing gas (initially in thermal equilibrium). The demon allows the passage of only fast-moving molecules from one vessel to the other and slow ones in the opposite direction. In this fashion, a temperature difference is established that could be used to extract work, as in a perpetual motion machine, in apparent violation of the second law of thermodynamics. In stochastic cooling, detection of the motion of groups of beam particles is used to correct their orbits in subsequent turns in circular accelerators; the net effect is the (irreversible) emittance reduction of the beam by the filling of internal 'voids' in phase space. In this process, the granularity of the beam plays a role in circumventing Liouville's theorem, which is based on a continuous fluid model.

An excellent introduction to stochastic cooling, invented by van der Meer in the early 1970s and first implemented at the CERN Intersecting Storage Rings (ISR), can be found in [15]. The basic scheme has a detector and a kicker separated by a quarter of the betatron wavelength plus integer multiples of the half-wavelength in a ring lattice. In this way, a particle error at the detector, where the betatron amplitude is near its maximum, yields a proportional signal that is sent to the kicker for correcting the orbit's slope. Thus, the orbit of the detected particle is corrected over successive turns, but the correction is not perfect mostly because of limited detector bandwidth, amplifier noise, and incoherent effects from the action of other particles. An *upper limit* to the (time) cooling rate can be estimated in a simple model that neglects incoherent effects and assumes ideal detection. If N is the number of particles in a ring, T the revolution period, and W the detector bandwidth, we can write $N_S = N/(2WT)$ for the number of particles per sample passing through the detector (we have assumed equally spaced samples of duration $1/2W$.) Further, the corrected kick is $\Delta x = -(\lambda N_S)\langle x \rangle_S$, where λ is a constant, and $\langle x \rangle_S$ is the average sample error. Therefore, with unit gain ($g = \lambda N_S = 1$) at the detector, the cooling rate is $1/\tau_n = 1/N_S$, *per turn*, and the net cooling rate is $(1/\tau_n) \times (1/T)$, in s^{-1}, or

$$\tau^{-1} = \frac{2W}{N}, \tag{3.6}$$

which overestimates the optimum cooling rate by a factor of 2. The cooling time is best for low-intensity beams. As an example [16], $N = 10^9$, $T = 0.5$ μs, $W = 250$ MHz for the Low-Energy Anti-proton Ring (LEAR), leading to $1/\tau = 0.5$ s^{-1}, i.e., a cooling time of 2 s, or 4 million turns.

Betatron oscillations and emittance of electron beams are damped from synchrotron radiation (SR) in circular machines (see appendix A). This damping also happens in very high-energy hadron machines such as the LHC. On the other hand, the standard stochastic cooling just described has been employed to improve the luminosity in proton–anti-proton machines (e.g., ISR) or heavy-ion colliders (e.g., RHIC), but it is impractical in electron rings or at the high energies and bunch densities of the LHC or Tevatron collider rings. At the LHC, for example [16], with a bunch rms length $\sigma_S = 9$ cm, and a circumference $C = 26.66$ km, the number of protons in the ring is $N = N_B C/\sigma_S$ where $N_B = 1.15 \times 10^{11}$ is the number of protons per bunch. Thus, the use of equation (3.6) with a bandwidth $W = 500$ MHz yields a cooling time of some 9500 h! Electron cooling (see above) is also unfavorable at the high energies of LHC.

Standard stochastic cooling relies on detectors with bandwidths in the radio-to-microwave frequency range, or about 100 MHz to 10 GHz. In contrast, effective stochastic cooling of electron and positron beams, or of high-energy, high-density hadron beams, requires bandwidths in the optical frequency range (10^{14} Hz or 100 THz). In OSC, undulators (see e.g., chapter 6 in [8]) are used for both detector pickup and kicker. The EM radiation from a particle in the pickup undulator is amplified and makes a correction kick to the same particle in the second undulator. A magnetic chicane is employed to delay the particle to give time for the amplified radiation to arrive at the kicker simultaneously with the particle. OSC is also possible without amplification of the radiation, in which case it is called passive or transit-time OSC. Schematics of the basic idea behind a passive OSC system are shown in figure 3.5. IOTA has demonstrated this type of OSC recently (2021) of its 100 MeV electron beam with one order of magnitude better cooling rate than the natural SR emittance damping [17]. 'Active' OSC, which involves optical amplification, is planned for 2023.

The 3 m OSC section is located between dipoles M4L and M4R, next to the RF cavity (see figures 3.1 and 3.2). Table 3.2 summarizes the parameters of the OSC experiment (see also table 3.1 for electron beam operation.) The OSC is off, and damping is entirely from SR, when the electrons and the undulator radiation at the kicker are longitudinally separated, i.e., for high and low bypass delays. The OSC

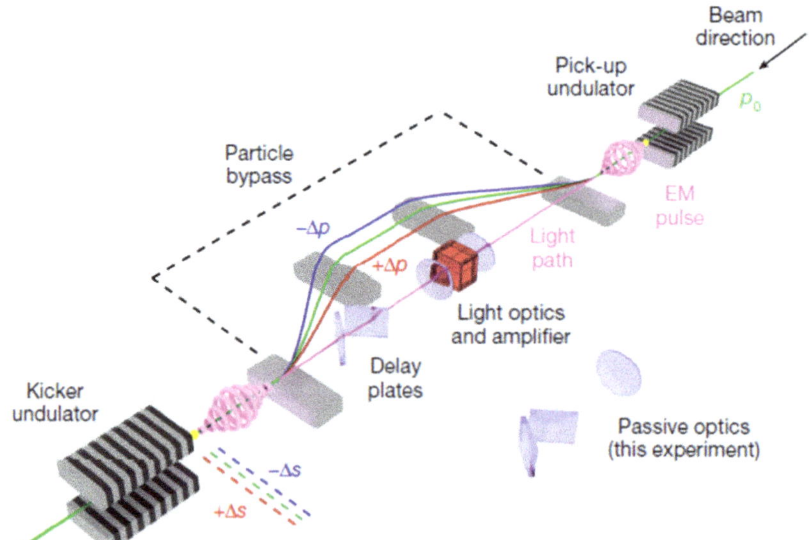

Figure 3.5. Schematics of a passive or transit-time OSC system. Trajectories of particles with three different momentum deviations are shown. Reproduced from [17], under the Creative Commons Attribution 4.0 International License http://creativecommons.org/licenses/by/4.0/.

Table 3.2. Partial list of parameters of passive OSC experiment in IOTA [17].

Design momentum, p_0	100 MeV/c
Momentum compaction factor	4.91×10^{-3}
Equilibrium RMS momentum spread (w/o OSC)	0.986×10^{-4}
Bypass delay	0.648 mm
Undulator radiation wavelength, λ_r	950 nm
Bandwidth of OSC system	19 THz
SR emittance damping rates	$(x,y,z) = (2.06, 0.94, 0.99)\ s^{-1}$

demonstrated in IOTA works for both transverse (betatron) and longitudinal (synchrotron) degrees of freedom, using techniques that may be applicable to other cooling methods. It requires stringent tolerances for beam/radiation alignment (100 μm, 100 μrad), and timing (sub-femtosecond). Further, low beam currents of 50–150 nA (10^5 particles, approximately) were used to reduce the effects of intra-beam scattering (IBS).

The section on 'Methods' in [17] describes in detail the diagnostics, instrumentation, and general techniques employed. Direct images of the beam transverse distribution are recorded with CMOS cameras as SR monitors at the main dipoles, while a streak camera is employed to image the longitudinal distribution. The measurements agree with theory after several corrections for diffraction and other effects are implemented. The 3D cooling rates, on the other hand, are calculated from changes in the equilibrium beam sizes in the longitudinal and vertical directions. A comprehensive discussion of the theory of the beam and light optics designs of the OSC experiment, as well as of simulations and instrumentation, appears in a recent paper by Lebedev *et al* [18]. However, a short analysis of the OSC rates is also presented in [17]. If **M** represents the 6 × 6 transfer matrix from pickup to kicker undulators, the longitudinal cooling rate for small-amplitude motion is given in the linear approximation by [17]

$$\tau_s^{-1} = \frac{\kappa}{2} f_0 k_0 (\mathbf{M}_{51} D + \mathbf{M}_{52} D' + \mathbf{M}_{56}), \tag{3.7}$$

where κ represents the maximum longitudinal kick, f_0 is the revolution frequency in the ring, k_0 is the wavenumber of the undulator radiation, and D is the dispersion function. For small fractional momentum deviations, the longitudinal kick is $\delta p/p_0 = \kappa u(s)\sin(k_0 s)$, where $u(s)$ is an envelope function and 's' is the longitudinal displacement of a particle relative to the reference particle in its motion from pickup to kicker undulators.

3.4 Transverse emittance measurements from undulator radiation noise [19]

Fluctuations in beam current arising from the discrete nature of the electrical charge is known as *Schottky noise*. Schottky detectors are employed in accelerators to measure parameters such as incoherent tune and chromaticity [20] and are key components of stochastic cooling. Fluctuations also occur in the number of photons from undulator radiation. In turn, the spectral content of this radiation is sensitive to bunch parameters. Thus, the variance in the photon number is given by

$$\mathrm{var}(N_{\mathrm{ph}}) = \langle N_{\mathrm{ph}} \rangle + \frac{\langle N_{\mathrm{ph}} \rangle^2}{M}, \tag{3.8}$$

where the linear term accounts for noise from the discreet (quantum) nature of radiation, and the quadratic term represents interference of the radiation from different electrons. The quantity M, known as the 'number of coherent modes', depends on bunch dimensions (and thus emittances) and rms momentum spread, as well as on parameters of the emitted radiation and properties of the detection system (e.g., efficiency).

In the experiment, electrons circulate clockwise. Radiation from an undulator located between M3R and M4R dipoles (see figure 3.2) exits at a window in MR4, and it is focused for detection by a photodiode in a dark box. The number of recorded photons at the ith turn in IOTA is proportional to a voltage amplitude signal A_i. To overcome limitations in digitization resolution, a signal splitter and a notch filter are employed to produce two signals, one proportional to $A_i - A_{i-1}$ (the Δ channel), and the other to $A_i + A_{i-1}$ (the Σ channel). Sub-millivolt turn-by-turn fluctuations are now measurable in the Δ channel. The mean and variance of N_{ph} are then calculated from recorded waveforms of Δ and Σ over 11 250 revolutions. When only one bunch parameter is unknown, it can be obtained from the measured var (N_{ph}) in combination with a numerical solution of a general formula for M. A detailed calculation of this parameter from first principles is presented in [21] where an electron beam with non-negligible divergence is assumed.

The method just described was applied to measure transverse emittances in two cases: coupled optics with a 'round beam,' and uncoupled optics with a 'flat beam.' The emittance of the round beam calculated from the standard method of using beam sizes and design Twiss parameters agrees with the fluctuations method. The horizontal emittance of the flat beam can be also obtained from the standard or fluctuations procedure; the vertical emittance, however, was too small to be calculable from signals from the synchrotron light monitors. The fluctuations approach, though, works for obtaining the vertical emittance of the flat beam: 5–15 nm, rms, unnormalized. Results of measurements for the flat beam are summarized in figure 3.6.

The nondestructive fluctuations-based diagnostics will work if var(N_{ph}) is dominated by the quadratic term, i.e., the one involving M, and if the latter is sensitive to transverse emittances. The technique can be useful as diagnostics for existing and next-generation low-emittance, high-brightness light sources.

3.5 Space charge compensation (SCC) and other experiments

Under the right conditions, an electron cloud superimposed on a proton beam can neutralize the effects of SC in the latter (see section 4.6 in [22]). The electron lens discussed above can be adapted to do SC compensation (SCC) in the IOTA

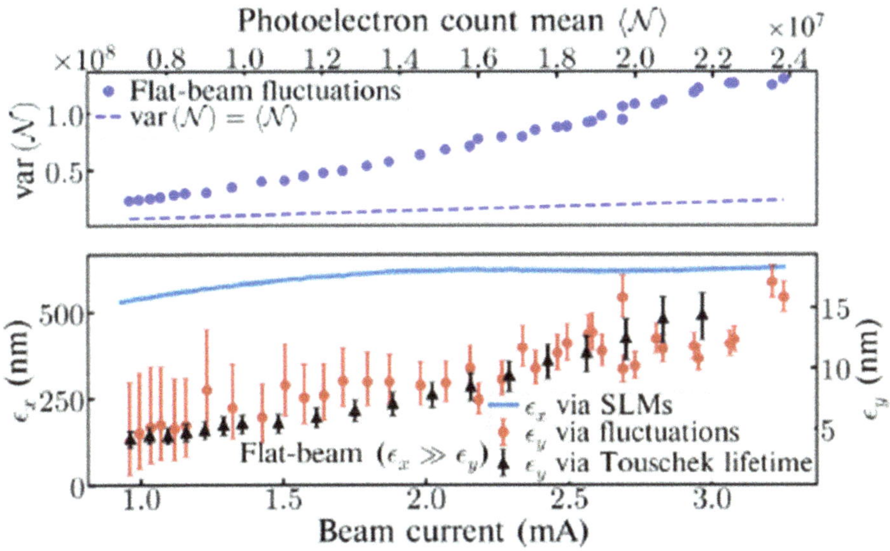

Figure 3.6. Variance of photon count for flat beam (top), and horizontal (bottom left scale, blue line), and vertical (bottom right scale) emittance data for flat beam. Reproduced from [19], under the Creative Commons Attribution 4.0 International License http://creativecommons.org/licenses/by/4.0/.

low-energy proton beam. There are two ways to accomplish this. In the first method, the electron gun generates the necessary charge distribution, transversely and longitudinally, to reproduce the bunch profiles of the circulating proton beam. In the second method, electrons are generated by ionization of the residual gas to create an 'electron column' that is contained radially by the main solenoid and longitudinally by electrodes, in a configuration similar to a Penning–Malmberg trap [23].

The transverse potential depression for a long charged-particle beam with uniform charge density is given by [22]

$$\phi = \frac{I_b}{4\pi\varepsilon_0\beta c} \cong 30\frac{I_b}{\beta}\ \text{(Volts)}, \tag{3.9}$$

where I_b is the beam current and β is the relativistic factor. We get 3.3 V for the 2.5 MeV, 8 mA IOTA proton beam. Thus, using a single 1 m electron column, the voltage on the electrodes for SCC of the circulating beam in the 40-m ring is estimated to be 130 V. Simulation studies with the WARP code (see appendix A in [8]) show that SCC can be optimized by varying the main solenoid field, electrode voltages, and vacuum pressure.

Furthermore, there are plans to do electron cooling of a 2.5 MeV, 0.44 mA, approximately, proton beam with an estimated cooling time of 20 ms and a reduction of transverse emittance by a factor of 10. It is of special interest to investigate how and if NIO leads to higher brightness when combined with electron cooling. There are also proposals to use the electron beam to generate x-rays, gamma rays, and tunable THz radiation, through several mechanisms, e.g., x-rays from diamond crystals, and gamma rays from inverse Compton scattering.

We conclude our summary of IOTA/FAST research with a very brief account of an experiment to detect a circulating single electron [5].

We quote from the Fermilab website on IOTA, 'On October 31 [2018], for the first time at Fermilab, we recorded a single electron traveling through an accelerator, the 40-meter-circumference IOTA ring. It circulated for four minutes, corresponding to about 2 billion turns. We set up sensitive diagnostics based on synchrotron radiation, which allowed us to observe individual electrons being lost due to collisions with the residual gas in the vacuum chamber. The very last electron died, poetically enough, on Halloween night at 22:41:25.' Figure 3.7 illustrates the process.

Additional experimental work with a single electron is described by Lobach *et al* in [25]. The study involved the statistics of undulator radiation and synchrotron motion of the electron. The use of a special binary detector confirmed that the photo-statistics follows a predicted Bernoulli-type of memoryless process. The investigation of electron motion, on the other hand, allowed the estimation of jitter in the phase of the RF cavity and the dependence of synchrotron period on amplitude.

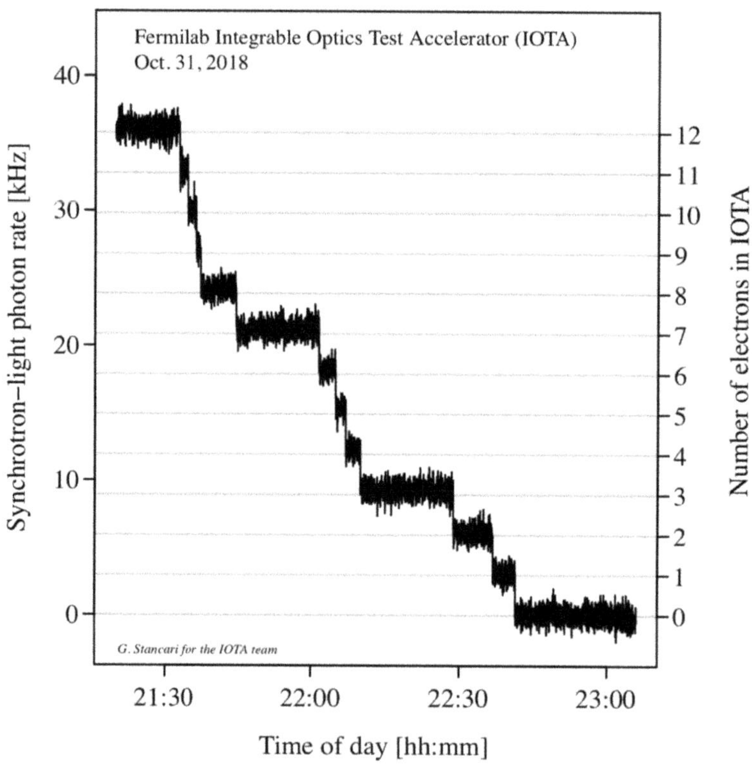

Figure 3.7. Detection of electrons in IOTA from emitted synchrotron radiation. Reproduced from [24]. Copyright © 2019 by JACoW — cc Creative Commons Attribution 3.0.

3.6 Prospects

The dual operation as a proton or electron machine makes IOTA unique among research accelerators. Moreover, early in the design of IOTA, it was realized that no existing accelerators would easily accommodate any of the envisioned projects, including nonlinear integrable optics lattices and OSC. With the demonstration of passive OSC for the first time in any accelerator, the realization of *active* OSC is only a matter of time. Progress in NIO is also guaranteed, as it is in other areas discussed above.

3.7 Computer resources

The author's website contains MATHCAD worksheets for IOTA's operation with either electrons or protons. The calculations aim to reproduce entries in the parameter tables (see text) such as incoherent and coherent SC tune shifts. Furthermore, the code OPTIMX from Fermilab comes with example files on IOTA (at least the Linux version) that reproduce the lattice functions and other calculations. See https://uspas.fnal.gov/resources/downloads.shtml.

References

[1] Antipov S *et al* 2017 IOTA (Integrable Optics Test Accelerator): facility and experimental beam physics program *J. Instrum.* **12** T03002

[2] Antipov S A *et al* 2017 Single-particle dynamics in a nonlinear accelerator lattice: attaining a large tune spread with octupoles in IOTA *J. Instrum.* **12** P04008

[3] Stancari G *et al* 2021 Beam physics research with the IOTA electron lens *J. Instrum.* **16** P05002

[4] Valishev A *et al* 2021 First results of the IOTA ring research at Fermilab *12th Int. Particle Acc. Conf. (IPAC Campinas, SP, Brazil)*

[5] Romanov A *et al* 2021 Experimental 3-dimensional tracking of the dynamics of a single electron in the Fermilab Integrable Optics Test Accelerator (IOTA) *J. Instrum.* **16** P12009

[6] Stancari G and Physics B 2022 Research at IOTA, presentation at the extreme rings workshop (ESRW22) *CERN*

[7] Ferrario M 2013 Space-charge dominated beams in guns and transport lines *Handbook of Accelerator Physics and Engineering* ed A W Chao and K H Mess (Singapore: World Scientific) 2nd edn

[8] Bernal S 2022 *A Practical Introduction to Beam Physics and Particle Accelerators* 3rd edn (Bristol: IOP Publishing)

[9] Hofmann A Landau damping *CERN Accelerator School Proc.*

[10] Alexander V *et al* 2020 Nonlinear integrable optics (NIO) in IOTA run 2 *Fermilab IOTA/FAST Experiment Report—Beams-doc-8871*

[11] Danilov V and Nagaitsev S 2010 Nonlinear accelerator lattices with one and two analytic invariants *Phys. Rev. ST Accel. Beams* **13** 084002

[12] Nagaitsev S, Valishev A and Danilov V 2010 Nonlinear optics as a path to high-intensity circular machines *Proc. of the 46th ICFA Advanced Beam Dynamics Workshop on High-Intensity and High-Brightness Hadron Beams, HB (Morschach, Switzerland)*

[13] Möhl D 2005 *Beam cooling CAS CERN Accelerator School 224–39* CERN-2005-004

[14] Sessler A M 1996 *Methods of Beam Cooling UC-414, Lawrence Berkeley Laboratory* (Berkeley, CA: University of California)

[15] Möhl D *Stochastic Cooling for Beginners CAS CERN Accelerator School 97–161* CERN-84-15

[16] Church M, Edwards H, Harms E and Henderson S *et al* 2012 *Proposal for an Accelerator R&D User Facility at Fermilab's Advanced Superconducting Test Accelerator (ASTA)*

[17] Jarvis J, Lebedev V and Romanov A *et al* 2022 Experimental demonstration of optical stochastic cooling *Nature* **608** 287–92

[18] Lebedev V *et al* 2021 The design of Optical Stochastic Cooling for IOTA *J. Instrum.* **16** T05002

[19] Ihar Lobach S N and Valeri Lebedev A R *et al* 2021 Transverse beam emittance measurement by undulator radiation power noise *Phys. Rev. Lett.* **126** 134802

[20] Caspers F 2008 *Schottky Signals for Longitudinal and Transverse Bunched-Beam Diagnostics* (Dourdan, France: CAS—CERN Accelerator School: Beam Diagnostics) pp 407

[21] Lobach I, Nagaitsev S, Lebedev V and Romanov A *et al* 2021 Measurements of undulator radiation power noise and comparison with *ab initio* calculations *Phys. Rev. Accel. Beams* **24** 040701

[22] Reiser M 2008 *Theory and Design of Charged Particle Beams* 2nd edn (Weinheim: Wiley-VCH)

[23] Dubin D H E and O'Neil T M 1999 Trapped non-neutral plasmas, liquids, and crystals (the thermal equilibrium states) *Rev. Mod. Phys.* **71** 1

[24] Nagaitsev S, Stancari G, Romanov A, Arodzero A, Murokh A, Ruelas M, Lobach I, and Shaftan T 2019 Experimental study of a single electron in a storage ring via undulator radiation, *10th Int. Particle Accelerator Conf., IPAC2019* pp 781

[25] Lobach I *et al* 2022 Single electron in a storage ring: a probe into the fundamental properties of synchrotron radiation and a powerful diagnostic tool *J. Instrum.* **17** P02014

Chapter 4

Fixed-field alternating gradient (FFAG) accelerators and the Electron Model for Many Applications (EMMA)

We describe in this chapter EMMA, a successful proof-of-principle machine as the world's first non-scaling fixed-field alternating gradient accelerator (FFAG). EMMA was in operation from 2010–2012 at the Daresbury Laboratory in the UK. We start in section 4.1 with a brief historical account and a description of scaling versus non-scaling FFAGs, giving a motivation for the *cardinal conditions* of the scaling machines. In section 4.2, we cover the layout and main parameters of EMMA, and in section 4.3, we summarize the main results of research in EMMA, i.e., integer resonance crossing via so-called serpentine acceleration over six turns without beam losses. In section 4.4, we describe the generalities of a muon–muon collider complex based mostly on the lucid *Physics Today* article by Sessler [1], including the main technical challenges and key components such as muon cooling. In section 4.5, we list important issues that could have been investigated in a continuation of research with EMMA, according to a 2014 presentation by Machida. We conclude by mentioning computer resources related to EMMA, cyclotrons and FFAGs in general.

4.1 Scaling versus non-scaling fixed-field alternating gradient accelerators

Particle accelerator science and technology has been dominated for decades by alternating gradient (AG) synchrotron machines, either for the acceleration of electrons or hadrons or as light sources. However, the limitations of synchrotron rings for new projects in high-energy physics, nuclear physics and applications in industry and medicine have prompted new ideas, or the revisiting of old ones, for more efficient and economically feasible machines. An example from high-energy

doi:10.1088/978-0-7503-4891-1ch4

physics is the muon–muon collider that requires very rapid acceleration of the fast-decaying muons (see [1] and additional references in section 4.4). EMMA, was designed as a scaled version of a muon accelerator. EMMA is the first linear non-scaling FFAG to be built and successfully tested. It was operational in Daresbury Laboratory in Cheshire, in the UK, from 2010–2012.

Symon and Kerst in the USA [2], Ohkawa in Japan, and Kolomensky in the former Soviet Union invented the FFAG concept in the mid-1950s, a few years after the invention of AG, or strong, focusing by Courant, Christofilos and others. Some of the shortcomings of the original Lawrence cyclotron had been overcome with the synchrocyclotron and the sector focused cyclotron, and the FFAG would use techniques of the new cyclotrons. See [3] for an introductory treatment and history of both cyclotrons and FFAGs.

Symon and collaborators from MURA (Midwestern Universities Research Association) in Wisconsin built several prototype electron FFAGs of the scaling type, but their proposals for proton machines were not approved and FFAGs were abandoned [4]. However, the machines were revisited in Japan in the late 1990s and early 2000s when several proton accelerators were built as proof-of-principle scaling FFAGs for nuclear physics and medical applications [5].

Interest in FFAGs is illustrated by the international workshops that have taken place since 1999, sometimes two per year, with the next one to be held in September 2023 (FFA'23) at Thomas Jefferson National Accelerator Facility in Newport News, VA. Some of these meetings, e.g., FFA'19 at the Paul Scherrer Institut (PSI) in Switzerland, included an 'FFA School', with lectures on the design of FFAGs using freely available codes. The last section on computer resources contains additional information.

As the name implies and unlike synchrotrons, FFAGs have magnets that are not ramped during acceleration. This allows faster repetition rates (limited only by RF capabilities), higher average beam current and larger momentum acceptance. However, the magnets must have large apertures, short lengths, and require special nonlinear B-field radial profiles. In the *scaling* FFAGs the magnets are designed so that the particle orbits and tunes are the same at all energies (zero chromaticity), thus avoiding resonance crossing on acceleration. Symon and collaborators [4] introduced two types of scaling FFAGs: *radial* and *spiral*. Figure 4.1 illustrates the basic design of a scaling FFAG of the radial type; it consists of eight triplet FDF cells.

To understand the physics of *scaling* FFAGs, we start with the Kerst–Serber equations of motion, originally derived for the betatron accelerator. Let us consider motion of a charged particle on the horizontal x–y plane under the influence of a vertical B-field. If the particle moves off the design orbit with a small offset x, so that $r = \rho_0 + x$, with $x \ll \rho_0$, we then obtain [7]

$$x''(s) + \frac{1}{\rho_0^{\,2}}(1 - n)x(s) = 0, \;\; z''(s) + \frac{n}{\rho_0^{\,2}}z(s) = 0. \tag{4.1}$$

where 'n' defines the *field index*:

$$n \equiv -\frac{\rho_0}{B_0}\left(\frac{\partial B_z}{\partial r}\right)_{r=\rho_0}. \tag{4.2}$$

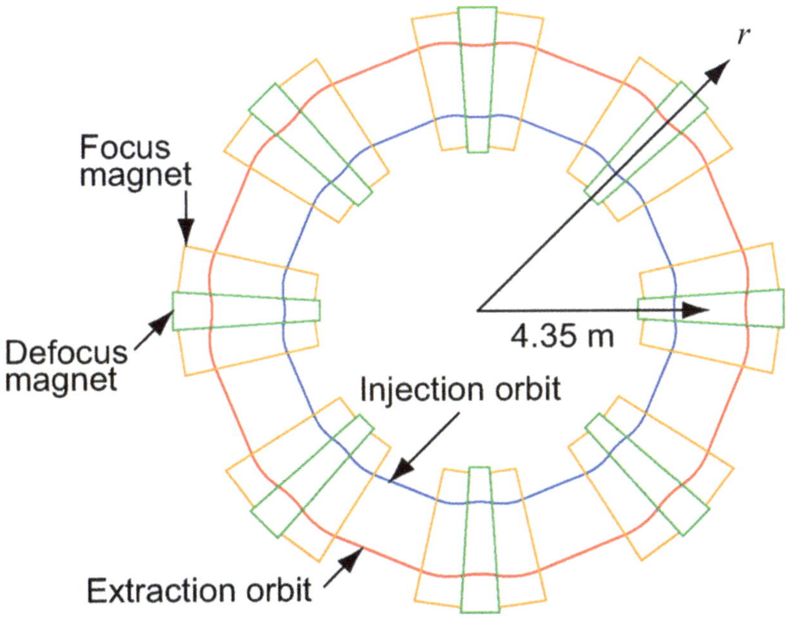

Figure 4.1. Schematics of a radial sector FFAG. Copyright (2012) IEEE, reproduced with permission [6].

In equation (4.2), ρ_0, B_0 represent equilibrium radius and vertical B-field at the magnet center. (We had already encountered 'n' in chapter 2 above equation (2.3) in connection with momentum compaction.) For particle stability, we require $0 < n < 1$, or $\nu_x = \sqrt{1-n}$, $\nu_z = \sqrt{n}$ in terms of horizontal and vertical *betatron tunes*.

It is convenient to rewrite equation (4.1) in the (x, θ, z) coordinate system, with $s = R\theta$:

$$\begin{aligned}\frac{d^2x}{d\theta^2} + \frac{R^2(\theta)}{\rho_0{}^2(R, \theta)}(1-n)x = 0,\\ \frac{d^2z}{d\theta^2} + \frac{R^2(\theta)}{\rho_0{}^2(R, \theta)}nz = 0.\end{aligned} \tag{4.3}$$

where R is the local trajectory radius relative to the *ring center*, and ρ_0 is the local curvature radius [10].

The two *cardinal conditions* in a scaling FFAG [8–10] state that (1) the betatron tune should be independent of momentum (i.e., a condition for *zero chromaticity*), and (2) the orbits should be *self-similar*. Mathematically we have,

$$\begin{aligned}\left.\frac{\partial n}{\partial p}\right|_{\theta=\text{const}} = 0,\\ \left.\frac{\partial(R/\rho_0)}{\partial p}\right|_{\theta=\text{const}} = 0.\end{aligned} \tag{4.4}$$

These conditions are satisfied if the *azimuthal average* of the *B*-field follows the profile [2, 9–11]

$$B_{av}(R) = B_0\left(\frac{R}{\rho_0}\right)^k, \tag{4.5}$$

where $B_{av}(R) = \langle B(\theta)\rangle$ is the azimuthal average over the median plane of the magnet, B_0 is the field at the magnet center, R is the radial distance from the FFAG center to the *equilibrium orbit*, and k is a constant, $k \gg 0$ defined as

$$k(R) \equiv \frac{R}{B_{av}}\frac{dB_{av}}{dR}, \tag{4.6}$$

where $B_{av} = \langle B(\theta)\rangle$ (compare with equation (4.2)). Further, to accomplish both horizontal and vertical orbit stability, a positive-bending radial-sector magnet of field B_F is followed by a shorter negative-bending one of field $-B_F$. To first order, the betatron tunes are given by [11]

$$\nu_r = \sqrt{1+k}, \ \nu_z = \sqrt{-k + F^2(1 + 2\,\mathrm{Tan}^2\,\varepsilon)}, \tag{4.7}$$

where $F^2 = \langle (B - B_{av})^2\rangle / B_{av}^2$ defines the square of the *magnetic flutter*. For radial scaling FFAGs the angle $\varepsilon = 0$, and the flutter is constant. AG focusing is also realized with edge focusing in the *spiral* type of scaling FFAGs. In these machines, particles enter the magnet faces at an angle ε relative to the normal (see section 2.5 of [7]), and no alternating polarities are used. To summarize, scaling FFAGs operate with constant tunes from the injection energy to the extracted one, i.e., zero chromaticity, and with self-similar orbits.

With fast acceleration (in less than 20 turns) in mind, Mills, Johnstone and collaborators proposed and designed FFAGs without scaling [12]. *Non-scaling* FFAGs allow the betatron tunes to vary, crossing multiple integer resonances during acceleration. Moreover, high periodicity is required to accommodate large momentum acceptance. However, there is more flexibility for lattice design, leading to larger DAs and smaller physical apertures. In the *linear* type of non-scaling FFAGs, or LNS-FFAGs, standard linear quadrupole magnets are used, and the orbit circumference varies quadratically with momentum enabling the use of fixed RF frequency at relativistic energies [11].

Near or at the minimum in the orbit circumference as a function of energy, LNS-FFAGs have very small compaction factors, i.e., the machines are nearly isochronous (see section 2.2 on the SIR machine). Furthermore, the Hamiltonian for longitudinal motion has a cubic dependence on energy [10] that leads to stable fixed buckets at two different energies in the energy versus phase diagram. Thus, it becomes possible to accelerate between buckets instead of inside them, leading to so-called 'serpentine' or 'gutter' acceleration as in nearly isochronous cyclotrons.

4.2 Layout and main parameters of EMMA

As already mentioned, EMMA is a proof-of-principle LNS-FFAG. It was designed to demonstrate fast and safe crossing of multiple resonances during acceleration, and serpentine acceleration. The main parameters of EMMA are summarized in table 4.1, and diagrams of the machine and main cell are displayed in figure 4.2. Figure 4.3 shows a photograph of the commissioned EMMA at Daresbury, UK.

EMMA's design is described in great, mostly engineering, detail in [15]. The definite research results appear in a 2012 *Nature Physics* article [13] that also earned the cover of the journal with a depiction of serpentine acceleration. A good summary of EMMA's results can be found also in a presentation by Machida for the 2014 FFAG workshop at Brookhaven National Laboratory [16]. The material that follows relies on the latter three sources.

EMMA's main design feature is the high density of magnets and diagnostics. It has 84 short quadrupole magnets that can be displaced horizontally to produce the dipole component needed for bending the beam around the ring. Thus, eight different lattices with specified F and D quadrupole strengths and displacements were designed. The eight lattices have different cell-tune footprints as a function of

Table 4.1. Nominal parameters of EMMA [11, 13, 14].

Energy, ***E***	10–20 MeV
Circumference	16.57 m
Cells (focusing–defocusing doublets)	42
F quadrupole length	5.88 cm
D quadrupole length	7.57 cm
Rev. time @ 18.5 MeV c^{-1}, equiv. momentum	55.3 ns
Radiofrequency	1.30 GHz
No. of cavities	19 × 120 kV
Tune shift per cell, ring	$\approx 0.3 - 0.1$, $\approx 12 - 4$
Normalized transverse acceptance	3 mm
Bunch charge	10 – 40 pC

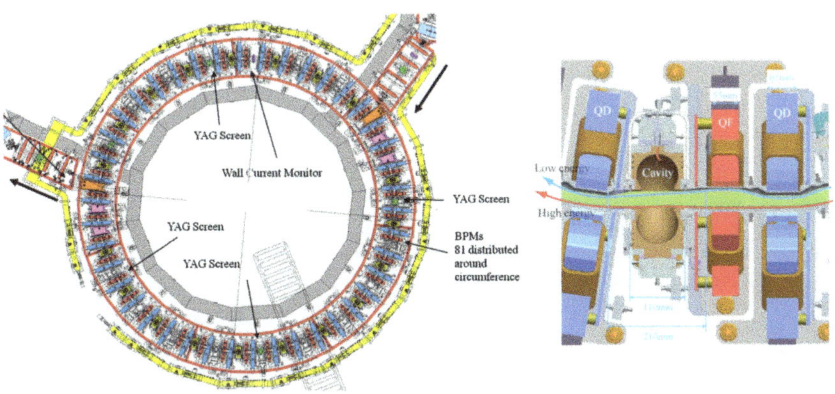

Figure 4.2. Schematics of EMMA and doublet cell. Reproduced from [15], copyright (2010), with permission from Elsevier.

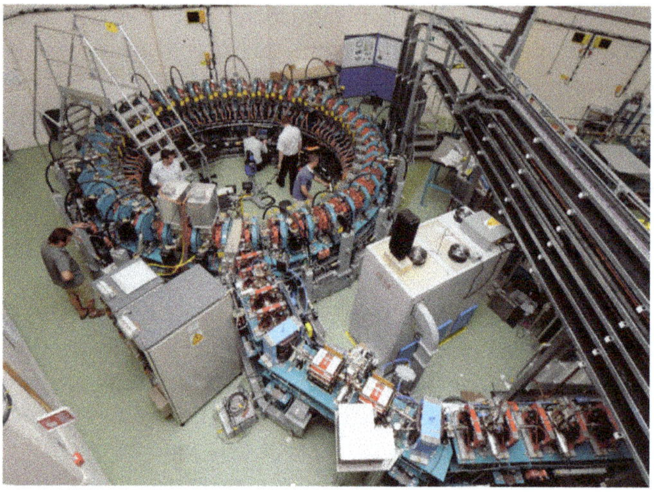

Figure 4.3. Photograph of EMMA at Daresbury, UK. Reproduced from [15], copyright (2010), with permission from Elsevier.

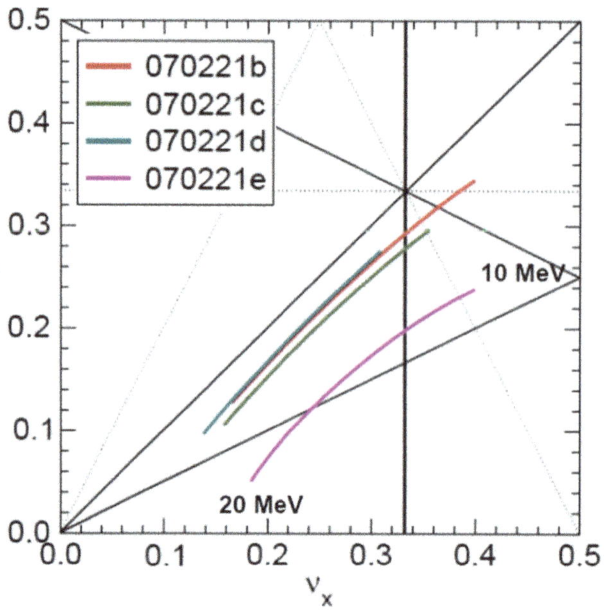

Figure 4.4. Cell tune footprint for four 'symmetric' lattices. Resonance lines up to third order are shown. Lattice labeled 'd' does not cross any resonance lines. Adapted from [11, 15]. Reproduced from [15], copyright (2010), with permission from Elsevier.

energy, as illustrated in figure 4.4 for four lattices (b–e) designated 'symmetric' in connection to the parabolic time of flight. Importantly, the quadrupoles in a family should have the same integrated field strength within 0.01%, according to simulations.

There is a total of 19 normal conducting, single cell RF cavities, one at every other section except at injection and extraction, with a baseline RF frequency of 1.30 GHz. Other than the obvious use of RF to accelerate the 10 MeV injected electrons, it is also adjusted for time-of-flight measurements at a fixed energy and for resonance crossing studies.

Diagnostics include 81 capacitive BPMs, two per cell between quadrupoles (except at injection, extraction), and a straight section with a WCM (figure 4.2). The overall resolution of the BPMs is 50 μm. There are also 16 YAG-crystal screens, four in the ring and six in both the injection and extraction beamlines. The latter beamlines have tomography sections for emittance measurements, consisting of three YAG-crystal screens and several quadrupole magnets. The injector comprises the ALICE (Accelerators and Lasers in Combined Experiments) facility: two linacs with superconducting cavities with acceleration capability up to 35 MeV. The horizontal acceptance of EMMA is probed with the two injection kickers. Particularly challenging were the large angles required for injection and extraction, 65° and 70°, respectively. Additional diagnostics in the original design included Faraday cups at the injection and extraction beamlines, a dipole spectrometer to measure beam energy on extraction, and an electro-optic device for longitudinal profiling of the beam.

4.3 Acceleration and integer-tune resonance crossing

Muratori reviews commissioning of EMMA in a detailed presentation at the 2011 FFAG Workshop at Daresbury [14]. The machine was doing more than 1000 turns without RF by August 2010. Acceleration from 12.5 to 21 MeV c^{-1} momenta was achieved in April 2011. At an RF voltage of 100 kV per cavity, i.e., a total of 1.9 MV for all cavities, a wide channel between RF buckets was created for so-called serpentine acceleration for several initial phase values. The orbit and cell-tune were obtained as a function of cell number for five consecutive turns. The results are shown in figures 4.5(a) and (b) for one case of initial RF phase.

The trajectories in longitudinal phase space were determined from beam position and tune measurements. The results, with momentum obtained from horizontal beam position, for five values of initial phase values are displayed in figure 4.5(c).

In conclusion, EMMA demonstrated serpentine acceleration within six turns from 12 to 18.4 MeV c^{-1}, equivalent momentum, with a tune variation greater than 4.2 (cell tunes are multiplied by 42—see figure 4.5(b)) and orbit excursions of the order of 10 mm. The latter figure is well within the 40 mm aperture of the D quadrupoles. Apparently, crossing of integer resonances causes coherent motion without emittance growth, but the momentum spread, and lack of chromaticity correction leads to decoherence and emittance growth [16].

4.4 Muon–muon collider

The Higgs boson is the carrier of the Higgs field, the field that gives mass to leptons and hadrons. It was proposed by Peter Higgs *et al* in 1964, and discovered at LHC in 2012 at an energy of 125 GeV. Sessler, on the other hand, had said in 1998 [1]: 'If m_{h}

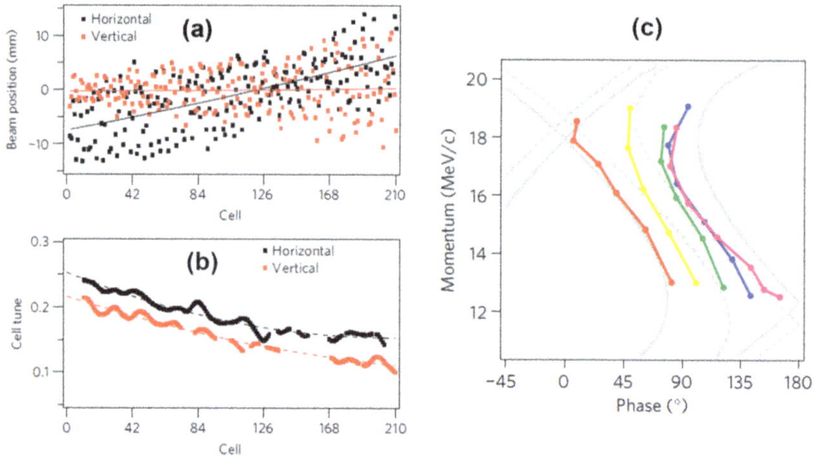

Figure 4.5. (a) Horizontal and vertical beam position for accelerated beam over five consecutive turns. (b) Horizontal and vertical cell tunes over five turns. (c) Acceleration in serpentine channel: trajectories in longitudinal phase space for five different values of initial RF phase. Reproduced from [13], copyright (2012), with permission from Springer Nature.

[the Higgs mass] turns out to be as little as 100 GeV, then a rather low-energy machine (say 50 GeV on 50 GeV), might be the first muon collider one seeks to build.' He also was very optimistic when he declared, 'looking broadly, and also contemplating what the US will do in high-energy physics, one may imagine a $\mu^+ \mu^-$ collider in the US, early in the next century [2000s?].' Twenty-five years after this statement, there has been renewed interest in a muon collider as a high priority international endeavor in R&D towards a machine in another 20 years or so. In fact, the International Muon Collider Collaboration (IMCC, https://muoncollider.web.cern.ch) has been established to coordinate the effort. Several workshops and seminars have taken place recently (April–May 2023) at Fermilab, Argonne National Lab, University of Maryland and elsewhere, on the physics and rationale for a muon collider as a high-energy frontier machine.

Because the muon mass is 207 times the electron mass, a muon circular collider would not suffer from the strong SR losses of an electron machine of similar energy (see appendix A, equations (A.11) and (A.12)). In addition, a collider based on leptons such as muons is much 'cleaner' than a hadron collider is because leptons are not composite particles as hadrons are. Thus, a 10 TeV muon collider would be competitive against a 100 TeV proton collider. The muon collider would be both a 'precision and a discovery' machine [17] capable of not only refined measurements of the Higgs particle parameters, but also of new findings.

However, there are many technical challenges associated with the muon fast decay, 2.2 μs at rest, or a few milliseconds for a TeV muon, and the decay products (neutrinos and electrons) which can be hazardous and complicate magnet and detector designs. Therefore, production, cooling, manipulation, and acceleration of muons must all be performed very fast, and special shielding at the detectors is needed.

The muon collider was originally proposed by Budker some 50 years ago and developed by Skrinsky in the former Soviet Union, and Neuffer in the USA. The main components of a muon collider complex are illustrated in figure 4.6: proton driver to produce pions, cooling section of muons, low and high-energy muon accelerators, collider ring and detection. FFAGs are among the accelerators being considered for this complex.

The first thorough study for a muon collider complex was completed in 2007 by B Palmer (Brookhaven National Lab) and collaborators, and the Muon Accelerator Program (MAP) was established in 2011 under the direction of M Palmer, also from BNL.

The *luminosity* (see appendix A, section A.4 for definition) required for a 3 TeV (collision energy) $\mu^+ \mu^-$ collider is 5×10^{34} s^{-1} cm^{-2}. The details of the collider design can be worked out backwards from this luminosity to determine the necessary proton beam intensity to produce pions. The muons resulting from the decay of pions are much disorganized, requiring cooling to produce a beam of reasonable *emittance*. The only cooling scheme that works fast enough for muons is *ionization cooling* (see chapter 3, section 3.3). In the process, the muons are 'heated' by scattering with the ionization material, resulting in an equilibrium, rms normalized emittance equal to 270 μm, approximately. The muon cooling has to be done in all three directions ('6D cooling') by a factor of about 100 in each direction. So far, ionization cooling of muons has been demonstrated in two dimensions ('4D cooling') in the MICE (muon ionization cooling experiment) project in the UK [18]. Muons are generated from a target in the ISIS synchrotron at the Rutherford Appleton Laboratory. Then, a muon beam with an initial high emittance (4–10 mm) and central momentum tuned between 140 and 240 MeV c^{-1} is cooled by a liquid

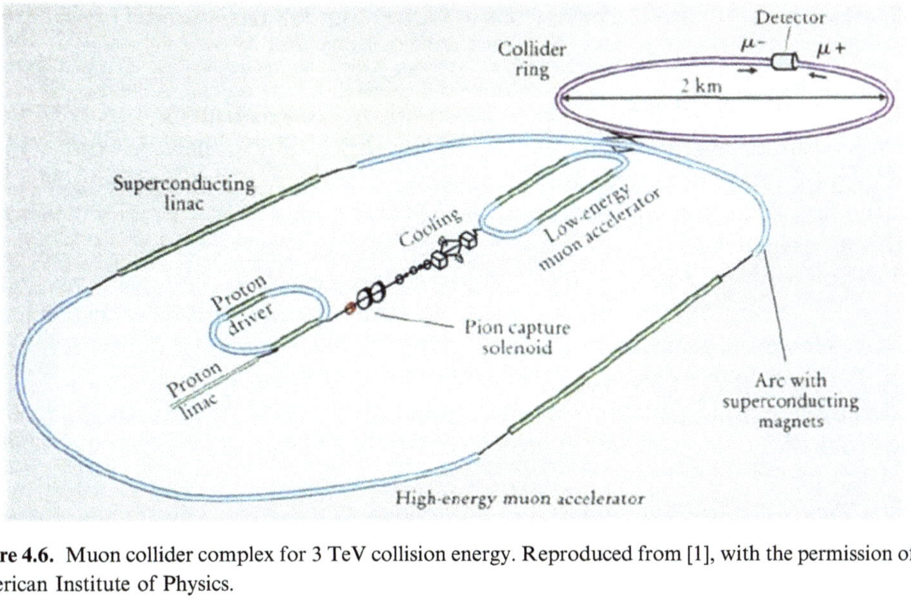

Figure 4.6. Muon collider complex for 3 TeV collision energy. Reproduced from [1], with the permission of the American Institute of Physics.

hydrogen or liquid lithium-hydride absorber. Cooling is demonstrated by an observed increase in the number of particles with small amplitudes measured from the beam center. Twelve superconducting solenoids (4T maximum field) are used in the MICE apparatus for tight focusing of the muon beam [18].

The need for muon cooling is avoided in another concept for a muon collider called LEMMA (low emittance muon accelerator) [17]. In LEMMA, muons are produced by the interaction of 45 GeV positrons with electrons; the resulting muons have low emittance, but very high positron currents are needed to compensate for the low probability of muon production. However, lower overall charge would be required in the collider (also allowing for higher energy), thus reducing the problems associated with the background from muon decay.

There are many proposals for the low energy and collider muon accelerators. According to many, however, NS-FFAGs provide the most scientifically and economically feasible scheme for a muon complex. Ruggiero, for example [19], summarizes a proposal for a three-ring solution for the low energy part, with the last ring being a 10–20 GeV, 440 m circumference NS-FFAG, the scaled-up version of EMMA (except possibly for the use of FDF triplet cells in the muon proposal as opposed to the EMMA doublets). The muon beam would circulate about 10 turns in each ring, would follow serpentine ('gutter') acceleration by means of superconducting RF cavities at 200 MHz, and be essentially isochronous. The latter feature makes it possible to have large momentum aperture, fixed RF frequency and short bunches. The momentum aperture, however, may be hindered by nonlinear resonances induced by space charge, as discussed by Lee [19, 20].

Another application of a muon machine would be as a *neutrino factory*. The muons would circulate in a storage ring and decay into equal numbers of electron and muon neutrinos. The energy of the neutrino beans would be tunable by varying the muon's energy. FFAG concepts for a 'decay ring' have been proposed in the nuSTORM project, with a clear synergy with muon collider studies [17].

4.5 EMMA prospects

'EMMA stopped its operation in December 2012. Since then, several funding requests have been submitted without success.' These are words by Shinji Machida from the first slide of his presentation on EMMA at the 2014 FFAG workshop at BNL [16]. Apparently, once EMMA achieved its main objective of serpentine acceleration with safe integer resonance crossing, and made the cover of *Nature Physics* [13], the funding agencies in the UK saw no reason for further research with EMMA. However, as in most research projects, there were significant unfinished issues. Perhaps, in the current (2023) environment of great enthusiasm for a muon collider, things would have turned out different for EMMA. We list below from [16] some of the areas that were left unanswered with EMMA, and which would likely be important in any actual NS-FFAG for a muon collider or other applications.

First, *beam-position monitor* (BPM) diagnostics, a critical component for orbit detection and control in any accelerator, required refined characterization in terms of accuracy and sensitivity when the beam reached near the aperture limit.

('Is there any quick fix to widen the detection area of BPM?') This concern is shared by UMER, but, apparently, not in IOTA (where EMMA's BPMs are now used) which has entirely different lattice and beam parameters (see chapter 3).

Second, lack of diagnostics made it difficult or impossible to ascertain orbit and optical mismatches *at injection*. 'Orbit in septum region [70° bend] not understood.' We encountered a similar situation in UMER, although the injection there is through the edge of a wide aperture pulsed quadrupole instead of a septum.

Third, and possibly related to the second item, it was difficult to steer the beam into the closed orbit below 12.5 MeV c^{-1}.

Fourth, 'explore optimized *muon lattice* configuration, namely QD/QF strength and position.'

Fifth, better magnet modelling with realistic fields, and simulation of entire ring. 'Online modelling with 3D field (ZGOUBI) of the whole ring (not hard edge model) will be nice.' The short EMMA magnets (figure 4.2, right) have fringe-dominated gradient profiles and there is field overlap between quadrupoles in a cell [21]. A valid hard edge model under these conditions may not be as straightforward to develop as for the short UMER quadrupoles (see chapter 1, and [7]) where there is no significant field overlap.

In conclusion, we do not make a bet for an EMMA II project any time soon, but as R&D for a muon collider advance, the need for similar FFAG electron model(s) will very likely arise.

4.6 Computer resources

The MATHCAD worksheet ***EMMA BasicCalc.xmcd*** (see author's website) contains basic calculations for the EMMA ring (table 4.1), including incoherent and coherent SC tune shifts. Given the 'bestiary of FFAG design studies' [10], the file ***MUON BasicCalc.xmcd*** is an attempt at calculations representative of a 10–20 GeV muon ring similar to the one mentioned by Ruggiero [19].

Calculations specific for FFAGs and cyclotrons can be done with the ray-tracing code ZGOUBI and the 3D space charge code OPAL-cycl. See appendix B and [22] for more information. Another option is the map code COSY INFINITY that has been updated with advanced capabilities for the modeling of complex FFAG structures [23].

References

[1] Sessler A M 1998 Gamma-ray colliders and muon colliders *Phys. Today* **51** 48

[2] Symon K R, Kerst D W, Jones L W, Laslett L J and Terwilliger T M 1956 Fixed-field alternating-gradient particle accelerators *Phys. Rev.* **103** 1837

[3] Craddock M K and Symon K R 2008 Cyclotrons and fixed-field alternating-gradient accelerators *Rev. Accel Sci. Technol.* **1** 65–97

[4] Symon K R 2003 Mura days *Proc. of the 2003 Particle Accelerator Conf., PAC03* 452

[5] Craddock M 2004 The rebirth of the FFAG *CERN Courier* https://cerncourier.com/a/the-rebirth-of-the-ffag/

[6] Keita Takahashi *et al* 2012 Magnetic field design of coil-dominated magnets wound with coated conductors *IEEE Trans. Appl. Supercond.* **22** 4901705
[7] Bernal S 2022 *A Practical Introduction to Beam Physics and Particle Accelerators* 3rd edn (Bristol: IOP Publishing)
[8] Machida S 2013 Fixed field alternating gradient *CERN Accelerator School Proc.* CERN 2013-001
[9] Lagrange J-B H R 2012 Study of Zero-Chromaticity in FFAG Accelerators *Dissertation* Kyoto University https://repository.kulib.kyoto-u.ac.jp/dspace/handle/2433/157577
[10] Méot F 2018 An introduction to fixed field alternating gradient accelerators *Presentation at Center for Accelerator Science and Education (CASE)* (Stony Brook University) http://case.physics.stonybrook.edu/images/2/2c/IntroSlides_FFAG.pdf
[11] Craddock M K 2011 FFAG OPTICS *Presentation at Int. Workshop on FFAG Accelerators (FFAG'11)* (Daresbury Laboratory)
[12] Johnstone C, Wan W and Garren A 1999 Fixed field circular accelerator designs *Proc. of the 1999 Particle Accelerator Conf., PAC99 (New York)*
[13] Machida S *et al* 2012 Acceleration in the linear non-scaling fixed-field alternating-gradient accelerator EMMA *Nat. Phys.* **8** 243–7
[14] Muratori B 2011 Status of EMMA *Presentation at Int. Workshop on FFAG Accelerators (FFAG'11)* (Daresbury Laboratory)
[15] Barlow R *et al* 2010 EMMA—the world's first non-scaling FFAG *Nucl. Instrum. Methods Phys. Res.* A **624** 1–19
[16] Machida S 2014 What we learned from EMMA *Presentation at Int. Workshop on FFAG Accelerators (FFAG'14)* (Brookhaven National Laboratory)
[17] Schulte D, Pastrone N and Long K 2020 Sketching out a muon collider *CERN Courier* **May–June** 41
[18] Muon Collaboration 2020 Demonstration of cooling by the muon ionization cooling experiment *Nature* **578** 53–9
[19] Ruggiero A G 2006 Challenges for hadron (and lepton) non-scaling FFAG *Proc. of HB (Tsukuba, Japan)* 303
[20] Lee S Y 2006 Fundamental limit of nonscaling fixed-field alternating-gradient accelerators *Phys. Rev. Lett.* **97** 104801
[21] Shepherd B J A and Marks N 2007 Quadrupole magnets for the 20 MeV FFAG, 'EMMA' *Proc. of PAC07, Albuquerque (New Mexico, USA)* 413
[22] Smirnov V 2017 Computer codes for beam dynamics analysis of cyclotronlike accelerators *Phys. Rev. Accel. Beams* **20** 124801
[23] Johnstonea C, Berz M, Makino K and Snopok P 2010 Isochronous (CW) non-scaling FFAGs: design and simulation *14th Advanced Accelerator Concepts Workshop 13–19 June 2010, Annapolis, MD, AIP Conf. Proc.* 1299 682–7

Chapter 5

Betatron resonances and space charge: Paul traps as model accelerators

We review in this chapter the use of linear Paul traps (LPTs) to simulate beam stability and betatron resonances in alternating gradient (AG) particle accelerator lattices. In the first section, we summarize the basic principles of LPTs and tabulate the parameters of three major simulator LPT experiments: the Paul Trap Simulator Experiment (PTSX) at the Princeton Plasma Physics Laboratory in the USA; the Simulator of Particle Orbit Dynamics (S-POD) at Hiroshima University in Japan; and the Intense Beam Experiment (IBEX) at the University of Oxford, and Rutherford Appleton Laboratory in the UK. We focus on PTSX research for the rest of the first section, establishing the connection between plasma and beam parameters.

In section 5.2, we describe research with S-POD as exemplified by a recent publication by Kojima *et al* in *Physical Review Accelerators and Beams*, and the controversy that it prompted with Hofmann from GSI in Germany.

In section 5.3, we focus on work with IBEX and additional studies with S-POD. The last two sections cover research prospects and computer resources.

5.1 Linear Paul trap (LPT) principles and PTSX research

A Paul trap, invented by Wolfgang Paul and collaborators around 1955, can confine ions in three dimensions by means of a transverse RF quadrupole field and axial DC fields. Okamoto *et al* in Japan (1999) and Davidson *et al* in the USA (2000), working independently, realized that the equations governing the dynamics of ions in a *linear Paul trap* (LPT) were isomorphic to those describing charged-particle beams in an AG lattice. In effect, 'the transverse RF quadrupole focusing in the laboratory frame (the time-domain) has the same property [character] as the discrete magnetic AG focusing in the beam frame (the s-domain)' [1]. Thus, an LPT can be used to model AG systems permitting the simulation of beams in almost any type of lattice

doi:10.1088/978-0-7503-4891-1ch5

geometry and at a very low cost. However, LPTs have some limitations: with fixed electrode lengths, modeling is restricted to 2D (no end effects), and no dispersive effects can be studied because of the lack of 'bending' elements.

Gilson, Davidson and collaborators presented results of experiments with PTSX (Paul Trap Simulator Experiment) at the Princeton Plasma Physics Laboratory in a 2004 *Physical Review Letters* publication [2], with additional details presented in a latter paper [3]. Okamoto and colleagues, on the other hand, worked extensively to simulate beam stability and resonances, with various degrees of SC intensity, using much smaller linear LPTs named S-POD (Simulator of Particle Orbit Dynamics) [4], later collaborating with groups in the UK (IBEX—Intense Beam Experiment). We illustrate PTSX and S-POD/IBEX in figure 5.1(a) and (b), and summarize their parameters in table 5.1. Figure 5.2 shows a photograph of S-POD.

If $\omega_0 = \omega_q$ represents the angular frequency of transverse focusing in a quadrupolar LPT, and f is the RF frequency, then the bare phase advance per RF cycle is $\sigma_0 = \omega_0/f$, and the corresponding betatron bare tune is $\nu_0 = \sigma_0/2\pi$. In the LPT [2, 7],

$$\omega_0 = \frac{8eV_0}{\pi m_b r_w^2 f}\xi, \quad \xi = \frac{\sqrt{2}}{4\pi}, \tag{5.1}$$

where the expression for ξ is valid for a *sinusoidal* RF waveform, V_0 is the voltage amplitude, and r_w is the trap radius. Therefore, the bare 'cell' tune in the case of a sinusoidal waveform is

$$\nu_0 = \frac{\sqrt{2}}{\pi^3}\frac{eV_0}{m_b}\left(\frac{1}{f\, r_w}\right)^2. \tag{5.2}$$

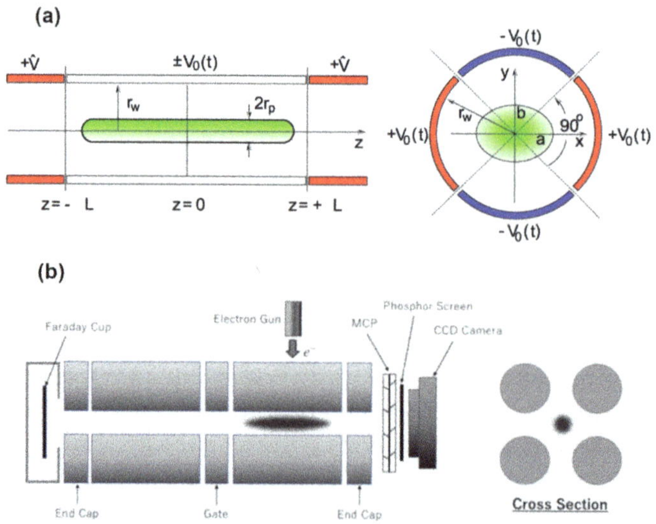

Figure 5.1. (a) PTSX device. (b) S-POD device. The IBEX trap is similar to S-POD. See table 5.1 for parameter values. Reproduced, with permission, from [2], copyright (2004) by the American Physical Society, and [4] CC BY 4.0.

Table 5.1. Main parameters of three LPT devices for accelerator physics research [3, 6].

	PTSX	S-POD II-III	IBEX
Length, $2L$	2.0 m	0.20 m	0.075 m
Ion species, m_b (amu)	C_s^+, 133	$^{40}Ar^+$, $^{40}Ca^+$	$^{40}Ar^+$
RF frequency, f	75 kHz	1 MHz	1 MHz
Voltage amplitude, V_0	235 V	< 100 V	< 100 V
Trap radius, r_w	10 cm	5 mm	5 mm
Typical plasma RMS radius, r_p	2.7 cm	1 mm	1 mm
Lowest tune depression, σ/σ_0	0.45	0.8	0.8
Temperature, k_BT	≈ 0.5 eV	≈ 0.25 eV	≈ 0.25 eV
Typical trapping time, t_t	> 100 ms	> 100 ms	> 100 ms

Gilson *et al* [2] use a 'normalized intensity parameter' $\hat{s} = \omega_p^2(0)/2\omega_q^2$ to quantify the SC intensity in PTSX. In the expression for $\hat{s}$, the numerator is the (on-axis) plasma frequency, squared, and the denominator contains the average transverse focusing angular frequency, squared, in the uniform-focusing approximation. It is straightforward to see that $\hat{s}$ corresponds exactly to the SC *intensity parameter* χ defined by Reiser [8, 9] as $\chi = 1 - \eta^2$, where η is the tune depression defined in equation (1.9) in chapter 1 (see also equation (1.2)). To see this, note that $K = \omega_p^2 a^2/2v^2$, $\omega_q^2 = \omega_0^2 = v^2 k_0^2$, also in the smooth approximation. Furthermore, the plasma 'global radial force-balance equation' can be translated into beam parameters. The plasma version, equation 4 in [2], is

$$m_b \omega_q^2 R_b^2 = 2k_B T + \frac{N_b e_b^2}{4\pi\varepsilon_0}, \tag{5.3}$$

with $R_b = a$, beam radius, k_B = Boltzmann constant (not to be confused with the wavenumber k), and N_b is the line particle density (in m^{-1}). The beam envelope equation in the smooth approximation was given in equation (1.3) (chapter 1) as

$$k_0^2 a - \frac{K}{a} = \frac{\tilde{\varepsilon}_x^2}{a^3}, \tag{5.4}$$

which can be seen to correspond to (plasma) equation (5.3) if we realize that $\omega_0^2 = v^2 k_0^2$, $\tilde{\varepsilon}_x^2 = k_B T a^2/m_b v^2$, $N_b = 2\pi a^2 n_b$, and use $\omega_p = q(n_b/2\varepsilon_0\gamma^3 m_b)^{1/2}$.

The line density N_b and plasma radius R_b are calculated from charge profile data obtained with a movable Faraday cup at the end of PTSX. Plotting the left-hand side of equation (5.3) versus the second term in the right-hand side yields a straight line for small values of N_b; thus, a transverse temperature $k_B T = 0.5$ eV is inferred. Large values of N_b, however, yield a departure from linearity, apparently because of a mismatch between the sizes of the ion source and the plasma. In any case, at an SC intensity of $\chi = 0.18$, no significant radial profile changes are observed in the plasma over trapping times of over 300 ms. This corresponds to 300×10^{-3} s $\times$ 75

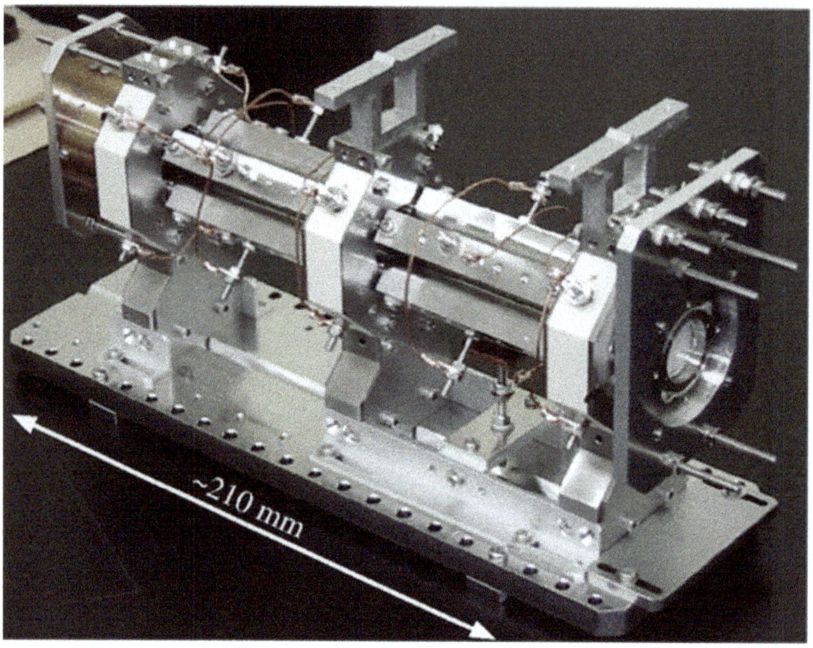

Figure 5.2. Photograph of linear Paul trap (LPT) S-POD. Reproduced from [5] CC BY 3.0.

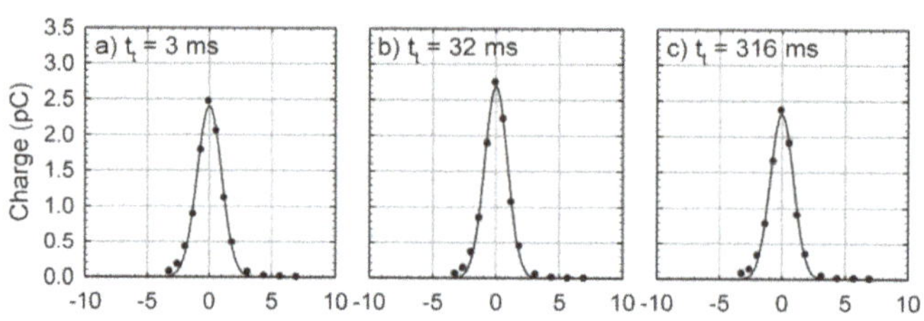

Figure 5.3. Evolution of radial profile of charge in PTSX for SC intensity $\chi = 0.18$. Reproduced, with permission, from [2], copyright (2004) by the American Physical Society.

kHz = 22.5k cycles, or 22.5 km, if the lattice full-period of the equivalent AG system is 1 m. Figure 5.3 illustrates the evolution of the radial profile.

Gilson *et al* [10] report observation of the envelope instability (see section 7.4 in [9]) at $\sigma_0 = 90°$, which can be obtained with $V_0 = 425$ V, $f = 75$ kHz (equation (5.2)) with sinusoidal trapping. Additional work with PTSX involved a study of beam degradation, as measured by on-axis charge decay in the plasma, induced by random noise applied to the trap voltage [11]. The noise was implemented in such a way as to mimic quadrupole gradient errors in an AG system; it was found that the transverse emittance increases in an approximately linear fashion with the noise duration, for noise levels up to 1.5% of voltage amplitude. The study was complemented with simulations with the code WARP (see appendix A in [9]).

5.2 S-POD (simulator of particle orbit dynamics)

Investigations with S-POD at Hiroshima University and with IBEX at the University of Oxford and Rutherford Appleton Laboratory have concentrated on experiments and simulations of betatron resonances with various SC intensities and lattice configurations (i.e., RF waveforms). The Hiroshima U. group has had a very good publication record on this complex subject over the last few years. Here, we illustrate some aspects of their work as presented in a recent paper by K Kojima, H Okamoto, and Y Tokashiki [12] and the interesting controversy that it triggered. The latter is represented by a comment from Hofmann [13], a well-known expert from GSI in Germany, and the reply by Kojima *et al* [14].

Kojima *et al* propose to use an expression similar to our equation (1.14) to empirically derive conditions for both 'coherent' (i.e., 'collective') and 'incoherent' (i.e., 'single-particle') resonances. In their notation we have,

$$k(\nu_{0x} - C_m \Delta\bar{\nu}_x) + l(\nu_{0y} - C_m \Delta\bar{\nu}_y) = \frac{n'}{2}. \tag{5.5}$$

In this equation, $\Delta\bar{\nu}_{x,\,y}$ represent *RMS betatron tune shifts*, C_m is the 'coherent mode coefficient'; k, l are integers such that $|k| + |l| = m$, the order of the resonance, and the RHS includes the extra factor of 1/2 that is associated with the envelope instability (see chapter 7 in [9]). Our first observation is that, apparently, Kojima and collaborators do not interpret the extra factor of 1/2 as being simply related to the envelope instability, but a general statement that in effect *doubles* the number of resonance stopbands for *all* orders m. The second observation is that the same semi-empirical coefficient C_m is employed for both transverse planes and it depends only on the order of the resonance. In their words, 'The semi-empirical rule proposed in [equation (5.5)] for the construction of a stability chart includes the incoherent aspect of resonances as a part. The two essentially different aspects, i.e., coherent resonances in the core and incoherent resonances in the tail, should suffice to describe the stability of the whole beam initially matched to the lattice.' [14]. Furthermore, 'the proposed semi-empirical formula in [equation (5.5)] enables one to plot resonance lines of practical importance very quickly and easily without solving a set of complicated differential equations numerically. It predicts the locations of instability bands under arbitrary initial beam conditions. All we need to know in advance are the basic lattice design plus only the rms tune depression that can readily be calculated from the rms envelope equations.' [14]. Figure 5.4 illustrates the concept for a revised tune chart.

Hofmann [13], on the other hand, does not think that the standard tune charts, which are based on the incoherent picture, should be revised. According to Hofmann, there is limited experimental evidence from real accelerators of second-order coherent resonances or instabilities, and no observations of higher than second-order coherent or parametric resonances.

There is also disagreement on the interpretation of solutions of Vlasov's equation [8] for coherent versus incoherent resonances. The 2019 Kojima *et al* paper asserts that Vlasov's theory precludes the existence of incoherent resonances in the beam

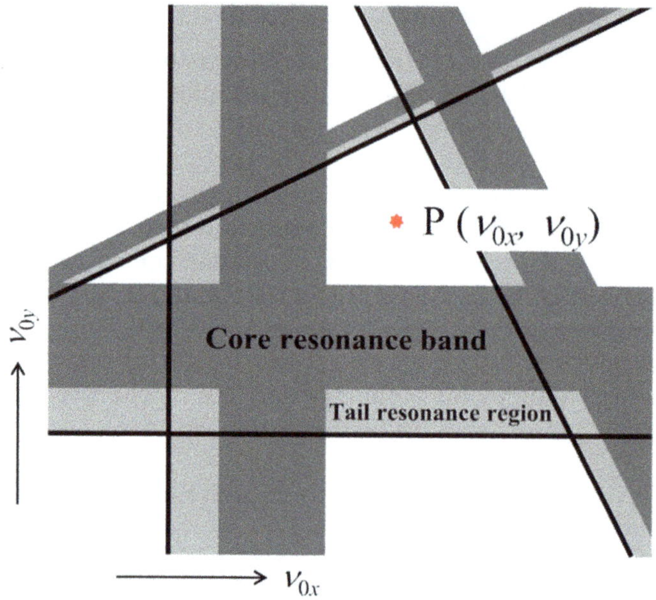

Figure 5.4. Conceptual illustration of desired operating point in bare-tune chart relative to the resonance stopbands derived from equation (5.5). Reproduced from [12] CC BY 4.0.

core, but Hofmann says that the incoherent spectrum is already taken out of the picture as part of a mathematical simplification towards solving Vlasov's equation. Furthermore, the doubling of stopbands (factor of 1/2 in equation (5.5)) is, according to Hofmann, valid only to second order. It would not work for higher orders because of *Landau damping* in the Gaussian-like beams treated by Kojima *et al.* Finally, experiments with GSI SIS-18 in Germany and with the proton synchrotron (PS) at CERN on third-order resonances (driven by sextupole errors) do not seem to agree with the empirical framework of Kojima *et al.* The latter defend their work by first emphasizing the simplicity and flexibility of S-POD experiments (and simulations with the code WARP), which have shown the existence of 'linear and low-order nonlinear coherent instabilities [and resonances].' As for comparison with experiments in real accelerators, Kojima *et al* present a case with the proton rapid cycling synchrotron (RCS) at J-PARC (Japan Proton Accelerator Research Complex) and reexamine the analysis of the high-intensity beam experiments in SIS-18.

5.3 IBEX (intense beam experiment)

To conclude, we mention work by the IBEX group in the UK [6], and additional research by the Hiroshima University group. Martin and collaborators study four different uncoupled resonances at cell bare tunes equal to $\nu_0 \equiv \nu_{0x} = \nu_{0y} = 1/8$, 1/6, 1/5, and 1/4, measure the RMS tune shifts, evaluate the A_m factors in the equation

$$\nu_0 - A_m \Delta \bar{\nu} = \frac{1}{2}\frac{n}{m}, \tag{5.6}$$

and compare the A_m's with the theoretical (from K–V distribution analysis) C_m's, equation (5.5) with $m = k = l$, and $n = n'$. The code WARP was used to simulate the experiments and evaluate RMS emittance growth over a small fraction (550) of the actual number of focusing periods in the experiment (10^5); Gaussian particle distributions were employed. As shown in figure 5.5, the resonances shift to higher bare tunes for increasing number of ions in the LPT. The tune shifts vary between 0.015 (at $\nu_0 = 1/4$) and 0.030 (at $\nu_0 = 1/8$) at the maximum ion number of 10^7; the corresponding SC intensity parameters (see equation (1.8) and definition of χ) vary between $\chi = 0.120$ and 0.480. Of particular interest is the resonance at $\nu_0 = 1/4$. A strong second-order parametric resonance (envelope instability), $n = 1$, $m = 2$ in equation (5.6), overlaps with the fourth-order incoherent resonance at $n = 2$, $m = 4$; the competition between these two had been studied in simulations by Hofmann and Boine-Frankenheim [15].

The slopes in plots of $\Delta\bar{\nu}_{\text{exp}}\,\varepsilon_{rms}$ versus trapped ion number N provide a measure of the coefficients A_m since we can write for the observed resonance shifts (figure 5.5) $\Delta\bar{\nu}_{\text{exp}} = A_m \Delta\bar{\nu}_{KV} \propto A_m N/\tilde{\varepsilon}_x$from equation (5.6) above, and equation (4.39) in [9] (note the difference between experimental and theoretical $\Delta\bar{\nu}$). The result is $A_m = 0.517 \pm 0.091$ for the 1/4 resonance, which, in our opinion, would correspond to the symmetric envelope mode discussed following equation (1.15) (chapter 1). However, Martin *et al* compare their result, for reasons that are unclear at the time of this writing, to the coefficient for the antisymmetric mode, i.e., 3/4. For the resonances at $\nu_0 = 1/8$, 1/6, on the other hand, the calculated coefficients A_m are larger but close to 1, as expected for 'incoherent' resonances. The case $\nu_0 = 1/5$

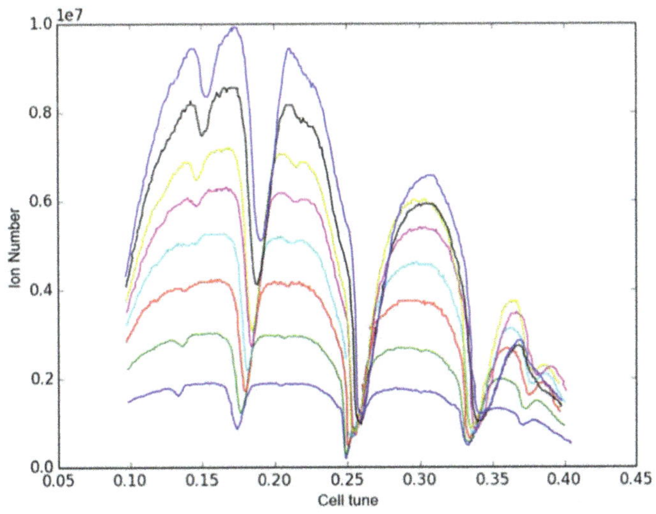

Figure 5.5. Extracted number of argon ions versus bare cell tune in experiments by the IBEX group. Each point represents one experiment. Reproduced from [6] CC BY 3.0.

yields $A_m \approx 0.8$, compared to the theoretical value of 1. No further analysis of this case is presented. The source of nonlinear resonances, i.e., $m > 2$, can be the nonlinear field components from the cylindrical electrodes of the LPT; an ideal quadrupolar LPT would use hyperbolic electrodes. Furthermore, some of the discrepancies between theory and experiment can be ascribed to the use of Gaussian or Gaussian-like particle distributions in experiments and WARP simulations versus calculations of C_m based on the K–V distribution. The resonances at $\nu_0 = 1/6$, 1/4, and other tunes have been studied also by the Hiroshima group [5, 16] and reviewed by Okamoto *et al* [4]. Their conclusion for $\nu_0 = 1/6$ is that it corresponds to an $m = 3$ mode (equation (5.6)) instead of a sixth order resonance; at $\nu_0 = 1/4$, beam quality degrades significantly even for fast resonance crossing. A scaling law for emittance growth is derived for the latter case.

5.4 Prospects

For their simplicity, flexibility, and compactness LPTs are elegant devices for modeling beam dynamics in particle accelerators with many different geometries of AG lattices. For example, the 42-fold symmetric lattice of the EMMA ring (figure 4.2) has been simulated in an LPT to study integer resonance crossing [5, 17–19]. Further, the 'ultimate control of tune depression' with laser cooling of $^{40}Ca^+$ ions in an LPT has been proposed to produce the equivalent of crystalline beams [18]. Other proposals include the construction of multipolar ion traps to excite nonlinear resonances in a controllable way [20], and to simulate IOTA's lattices with quasi-integrable nonlinear optics [21, 22]. All these ideas prove that LPTs and NL-PTs have a bright future in basic research of beam dynamics in accelerators.

5.5 Computer resources

The MATHCAD worksheet ***LPTs-BasicCalc.xmcd*** deals with basic calculations related to table 5.1 and equations (5.1)–(5.3), for all three LPTs covered in this chapter (PTSX, S-POD and IBEX).

WARP, a hybrid plasma/accelerator particle-in-cell code, is briefly described in Appendix A of [9]. Detailed information for installation and use, including examples, can be found in http://warp.lbl.gov/.

References

[1] Okamoto H, Endo M, Fukushima K, Higaki H, Itoa K, Moriya K, Yamaguchi S and Lund S M 2014 Experimental simulation of beam propagation over long path lengths using radio-frequency and magnetic traps *Nucl. Instrum. Methods Phys. Res.* A **733** 119–28

[2] Gilson E P, Davidson R C, Efthimion P C and Majeski R 2004 Paul trap simulator experiment to model intense-beam propagation in alternating-gradient transport systems *Phys. Rev. Lett.* **92** 155002

[3] Gilson E P, Chung M, Davidson R C, Efthimion P C, Majeski R and Startsev E A 2005 Simulation of long-distance beam propagation in the Paul trap simulator experiment *Nucl. Instrum. Methods Phys. Res.* A **544** 171–8

[4] Okamoto H, Higaki H, Ito K, Kojima K, Matsuba M and Tokashiki Y 2021 Application of a compact ion trap to fundamental beam-dynamics studies *Proc. 10th Int. Conf. Nuclear Physics at Storage Rings (STORI'17), JPS Conf. Proc.* 35 011007

[5] Takeuchi H, Fukushima K, Ito K, Moriya K, Okamoto H and Sugimoto H 2012 Experimental study of resonance crossing with a Paul trap *Phys. Rev. ST Accel. Beams* **15** 074201

[6] Martin L K *et al* 2019 A study of coherent and incoherent resonances in high intensity beams using a linear Paul trap *New J. Phys.* **21** 053023

[7] Davidson R C, Qin H and Gennady Shvets 2000 A Paul trap configuration to simulate intense non-neutral beam propagation over large distances through a periodic focusing quadrupole magnetic field *Phys. Plasmas* **7** 1020

[8] Reiser M 2008 *Theory and Design of Charged Particle Beams* 2nd edn (Weinheim: Wiley-VCH)

[9] Bernal S 2022 *A Practical Introduction to Beam Physics and Particle Accelerators* 3rd edn (Bristol: IOP Publishing)

[10] Gilson E P, Davidson R C, Efthimion P C, Majeski R and Qin H 2003 Recent results from the Paul trap simulator experiment *Proc. of the 2003 Particle Accelerator Conf., PAC03 (Portland, OR)*

[11] Chung M, Gilson E P, Davidson R C, Efthimion P C and Majeski R 2009 Use of a linear paul trap to study random noise-induced beam degradation in high-intensity accelerators *Phys. Rev. Lett.* **102** 145003

[12] Kojima K, Okamoto H and Tokashiki Y 2019 Empirical condition of betatron resonances with space charge *Phys. Rev. Accel. Beams* **22** 074201

[13] Hofmann I 2020 Comment on 'Empirical condition of betatron resonances with space charge' *Phys. Rev. Accel. Beams* **23** 028001

[14] Kojima K, Okamoto H and Tokashiki Y 2020 Reply to 'Comment on 'Empirical condition of betatron resonances with space charge'' *Phys. Rev. Accel. Beams* **23** 028002

[15] Hofmann I and Boine-Frankenheim O 2015 Space-charge structural instabilities and resonances in high-intensity beams *Phys. Rev. Lett.* **115** 204802

[16] Ito K, Okamoto H, Tokashiki Y and Fukushima K 2017 Coherent resonance stop bands in alternating gradient beam transport *Phys. Rev. Accel. Beams* **20** 064201

[17] Moriya K, Fukushima K, Ito K, Okano T, Okamoto H, Sheehy S L, Kelliher D J, Machida S and Prior C R 2015 Experimental study of integer resonance crossing in a non-scaling fixed field alternating gradient accelerator with a Paul ion trap *Phys. Rev. ST Accel. Beams* **18** 034001

[18] Okamoto H, Fukushima K, Higaki H, Ishikawa D, Ito K, Iwai T, Moriya K, Okano T, Osaki K and Yamaguchi M 2014 Beam dynamics studies with non-neutral plasma traps *Proc. 5th Int. Particle Accelerator Conf. IPAC (Dresden, Germany)* 4052

[19] Kelliher D J, Machida S, Prior C R, Sheehy S L, Fukushima K, Ito K, Moriya K, Okamoto H and Okano T 2014 Study of resonance crossing in non-scaling FFAGs using the S-POD linear Paul trap *Proc. 5th Int. Particle Accelerator Conf. IPAC (Dresden, Germany)* 1571

[20] Fukushima K and Okamoto H 2015 Design study of a multipole ion trap for beam physics applications *Plasma Fusion Res.* **10** 1401081

[21] Martin L K *et al* 2019 *J. Phys. Conf. Ser.* **1350** 012132

[22] Martin L 2019 Progress towards testing non-linear integrable optics in a Paul trap *Presentation at the 4th ICFA Mini Workshop on Space Charge* (CERN)

Appendix A

Physics scaling of particle ring accelerators

We summarize in this appendix a few basic concepts of ring accelerators. In the first section, we introduce basic particle dynamic parameters (magnetic rigidity, momentum, and cyclotron frequency) and relations from Special Relativity. In section A.2, we review formulae for synchrotron radiation (SR) (Larmor's formulae and related relations) and light sources, including radiation integrals, damping times and constants, equilibrium emittance and energy spread. In section A.3, we give a brief account of light sources, their classification and examples, and in section A.4 we discuss the basic physics of ring colliders with examples. The last section mentions computer resources available in the author's website.

A.1 Magnetic rigidity, relativistic momentum and cyclotron frequency

A charged particle moving in the presence of a uniform B-field follows a circular trajectory. By equating the Lorentz force, $\mathbf{F} = qv \times \mathbf{B}$, for a particle of charge q and velocity $v = \beta c\hat{\boldsymbol{\theta}}$ (c = speed of light in vacuum, $\hat{\boldsymbol{\theta}}$ = unit vector in the azimuthal direction), to the centripetal force we obtain

$$qvB = \frac{\gamma m v^2}{\rho}, \tag{A.1}$$

where ρ is the radius of the circular orbit, m is the rest mass, and γ is the relativistic mass factor $\gamma = (1 - \beta^2)^{-1}$. Therefore, with the relativistic linear momentum defined as $p = \gamma m V$, we have

$$B\rho = \frac{p}{q}. \tag{A.2}$$

The quantity '$B\rho$' is called the *magnetic rigidity*, and its units are Tm (Tesla-meter, in SI units), or Gcm (Gauss-cm, in CGS units). The reason for that name is that particles with higher energies and therefore higher '$B\rho$' are harder to 'bend': a higher

doi:10.1088/978-0-7503-4891-1ch6

field would be required for a fixed orbit radius ρ. It is strange (and sometimes confusing) to name the explicit product of two quantities with a single name, but it is almost universal usage in the accelerator-related literature. Perhaps only Wollnik [1] uses a single symbol (χ_B) for magnetic rigidity.

From the special theory of relativity, the relativistic momentum, total energy E and rest mass energy $E_0 = mc^2$ are connected by the relation

$$E^2 = p^2c^2 + E_0^2 \approx p^2c^2. \tag{A.3}$$

The last equality is valid for highly relativistic particles, i.e., particles for which $E \gg E_0$. For reference, $E_0 = 0.511$ MeV for the electron, $E_0 = 938$ MeV, or 0.938 GeV, for the proton, and $E_0 = 105.7$ MeV for the muon.

As an example, if B is given in Tesla (T) units and ρ in meters, we can write the following relation for *relativistic electrons*:

$$B[\mathrm{T}]\rho[\mathrm{m}] = 0.3p[\mathrm{GeV/c}]c, \tag{A.4}$$

where the momentum p is given in units of GeV/c and c is the speed of light. Thus, a 300 MeV electron will move on a circular trajectory of 1 m radius in a 1 T B-field. As an exercise, the reader can write the corresponding relation for the magnetic rigidity of relativistic protons and muons.

Finally, by writing $v = \rho(d\theta/dt) = \rho\omega_C$, we obtain from equation (A.1) the expression for the angular frequency of revolution of a particle of charge q in a uniform field B_0. It is called the *cyclotron frequency*:

$$\omega_C = \frac{qB_0}{\gamma m}. \tag{A.5}$$

A related frequency is the *Larmor frequency*:

$$\omega_L = \frac{\omega_C}{2}. \tag{A.6}$$

A.2 Synchrotron radiation [2]

The force that accelerates a charged particle can be resolved into two components: one along and another one perpendicular to the instantaneous velocity direction. The total power radiated by a charged particle is greater from transverse than from longitudinal acceleration by the relativistic factor γ^2:

$$P_\perp = \gamma^2 P_\parallel. \tag{A.7}$$

For highly relativistic particles, $\gamma \gg 1$; electrons with a kinetic energy of 10 MeV, for example, have $\gamma^2 = 423$. Thus, to generate copious EM radiation, it is more efficient to bend the trajectory of an energetic charged particle than to simply push it along a straight line. The radiation generated from transverse acceleration is called *synchrotron radiation (SR)*.

In the *non-relativistic limit*, i.e., when the speed is a small fraction of the speed of light c, the total instantaneous power (in watts) radiated by an electron (electrical charge—e) with acceleration $a(t)$ is given by *Larmor's formula*:

$$P_{\mathrm{NR}}(t) = \frac{e^2 a^2(t)}{6\pi\varepsilon_0 c^3}. \tag{A.8}$$

If relativistic effects are included, Larmor's formula becomes

$$P_{\mathrm{R}}(t) = \frac{e^2 a^2(t)\gamma^4}{6\pi\varepsilon_0 c^3} = P_{\mathrm{NR}}(t)\gamma^4 \tag{A.9}$$

If an electron of total energy $E = \gamma m_e c^2$ moves on a circular orbit of radius ρ under the action of a uniform magnetic field B, equation (A.9) can be written in other useful forms:

$$P_{\mathrm{R}} = \frac{e^2 c}{6\pi\varepsilon_0}\frac{\beta^4\gamma^4}{\rho^2} = \frac{e^4\beta^2}{6\pi\varepsilon_0 m_e{}^4 c^5}E^2B^2. \tag{A.10}$$

Equation (A.10) displays the scaling of SR power with *bending radius* ($P \propto 1/\rho^2$), *total energy* ($P \propto E^2$, β is essentially equal to 1 for highly relativistic particles), *rest mass* ($P \propto 1/m^4$), and *magnetic field* of bending dipole ($P \propto B^2$). Especially noteworthy is the very strong dependence of SR power on the rest mass of the charged particle. The energy lost to SR per turn for *one electron* of total energy E (in GeV) in a uniform field B (in Tesla) can be found from equation (A.10) after multiplying the power by $2\pi\rho/\beta c$, i.e., the revolution period:

$$\frac{\Delta E}{\mathrm{turn}}[\mathrm{keV}] = 88.4 \times \frac{E^4}{\rho} = 26.5 \times E^3 B, \tag{A.11}$$

where the bending radius ρ is given in meters. This energy must be replenished by RF accelerating cavities (see [2]) to keep the electrons circulating with the same radius ρ. From equation (A.10), we can see that the right-hand side of the first equality in equation (A.11) must be multiplied by $(m_e/m_p)^4 = 0.88 \times 10^{-13}$ to apply it to protons of the same energy and orbit radius as the electrons. This shows the limitations of electron circular machines for achieving high energy. For *muons*, the equivalent of equation (A.11) is

$$\frac{\Delta E}{\mathrm{turn}}[\mathrm{keV}] = 4.76 \times 10^{-8} \times \frac{E^4}{\rho} = 1.43 \times 10^{-8} \times E^3 B, \tag{A.12}$$

However, SR is in itself the reason for building electron machines as 'light sources'. The most common SR light source is the *electron storage ring*. The main lattice components in a storage ring are dipole magnets for bending and steering, quadrupole magnets for focusing, and sextupole magnets for chromaticity corrections. Thus, the storage ring lattice is characterized by the local radius of curvature $\rho(s)$ of the reference orbit, the quadrupole focusing function $\kappa(s)$, the horizontal betatron function $\beta_x(s)$, and the dispersion function $D_{0x}(s)$. The momentum compaction factor α_c, introduced in chapter 2 (equation (2.1)), the energy loss per turn, and other important storage ring parameters can be expressed in terms of integrals that were introduced by Helm *et al* in 1973. The first three *SR integrals* are given by

$$I_1 = \oint \frac{D_{0x}(s)}{\rho(s)} ds, \quad I_2 = \oint \frac{ds}{\rho^2(s)}, \quad I_3 = \oint \frac{ds}{|\rho^3(s)|}, \tag{A.13}$$

The momentum compaction factor as written in equation (2.16) is then

$$\alpha_c = \frac{I_1}{C_0}, \tag{A.14}$$

where C_0 is the length of the reference orbit. The energy loss per turn is,

$$U_0 = \frac{C_\gamma}{2\pi} E^4 I_2, \tag{A.15}$$

where

$$C_\gamma \equiv \frac{4\pi r_e}{3(m_e c^2)^3} \cong 8.844 \times 10^{-5} \frac{\text{m}}{\text{GeV}^3}. \tag{A.16}$$

The *nominal energy* is E, and r_e is the classical electron radius. Equation (A.15) generalizes the result in equation (A.10). The fourth SR integral is:

$$I_4 = \oint \frac{D_{0x}(s)}{\rho^3(s)} [1 + 2\rho^2(s)\kappa_x(s)] ds, \tag{A.17}$$

which is valid for a lattice made of sector magnets; for rectangular magnets (see section 2.5 in [2]), the factor in square brackets is reduced to $2\rho^2(s)\kappa_x(s)$. The fifth SR integral is

$$I_5 = \oint \frac{H(s)}{|\rho^3(s)|} ds, \quad H(s) = \beta_x D'^2_{0x} + 2\alpha_x D_{0x} D'_{0x} + \gamma_x D^2_{0x}. \tag{A.18}$$

The fourth and fifth SR integrals are used in expressions for the *damping constants* α_i $(i = x, y, z)$, stationary or 'natural' *rms energy spread*, $(\Delta E)_{\text{rms}}$, and *equilibrium transverse beam emittance*, ε_x, discussed next.

The emission of SR photons has important effects on the beam dynamics in storage rings and other SR sources. The first effect is the *radiation damping of betatron oscillations.* Because SR photons are emitted in a direction that is essentially identical to the direction of motion (i.e., tangential to the orbit) for highly relativistic charged particles, conservation of momentum of particle + radiation leads to an overall decrease of momentum of the particle. However, because RF replenishes only the momentum lost along the direction of the beamline, the vertical component of momentum is reduced after acceleration. Simple considerations and algebra (see, e.g., USPAS notes by Henderson *et al* [3] or Emery [4]) yield the following equation for the vertical component of the trajectory:

$$y'' + y'\frac{1}{E_0}\frac{dE}{ds} + \kappa_y y = 0. \tag{A.19}$$

The derivation depends on equating the fractional energy and fractional momentum gains: $dE/E = dp_{\text{photon}}/p_z$, valid with $\beta_0 = 1$ for highly relativistic particles. Equation (A.19) has the form of a damped harmonic oscillator equation; after changing from spatial to time derivatives, we find a *damping time* and *damping constant for vertical oscillations* equal to:

$$\tau_y = \frac{2E}{U_0}T_0, \ \alpha_y = \frac{1}{\tau_y}, \tag{A.20}$$

where U_0 is given by equation (A.15), and T_0 is the revolution period. The damping of betatron oscillations occurs in all three dimensions and goes along with a reduction in corresponding emittances. Thus, without considering other effects, the vertical emittance evolves like $\varepsilon_y = \varepsilon_{y0}\exp(-\alpha_y t)$. The damping times and constants for the components of motion in the *horizontal and longitudinal directions* can be similarly derived [5, 6]:

$$\tau_x = \frac{2E}{J_x U_0}T_0, \ \alpha_x = \frac{1}{\tau_x}, \tag{A.21}$$

$$\tau_z = \frac{2E}{J_z U_0}T_0, \ \alpha_z = \frac{1}{\tau_z}. \tag{A.22}$$

Equation (A.22) corresponds to the damping parameters of *synchrotron or energy oscillations*. The parameters J_i are called the *damping partition functions* and are equal to:

$$J_x = 1 - \frac{I_4}{I_2}, \ J_z = 2 + \frac{I_4}{I_2}, \tag{A.23}$$

where I_2 and I_4 are the SR integrals defined before.

The damping of betatron oscillations and emittance cannot go on without limit, as it is accompanied by an 'anti-damping' effect caused by random *quantum excitation*. Since the emission of SR photons also implies a reduction in energy, the particle finds itself moving very rapidly along different closed orbits. This orbit 'noise' leads to emittance increase but, because of the opposite effect from damping, a balance is eventually reached. The *equilibrium ('natural') horizontal emittance* and the *equilibrium ('natural') rms energy spread* are found to be [5, 6]:

$$\varepsilon_x = C_q\gamma^2\frac{I_5}{I_2 - I_4}, \ (\Delta E)_{\text{rms}} = C_q^{1/2}\gamma\left(\frac{I_3}{2I_2 + I_4}\right)^{1/2}. \tag{A.24}$$

The 'quantum constant' C_q in equations (A.24) is

$$C_q = \frac{55}{32\sqrt{3}}\frac{h}{2\pi m_e c} \cong 3.83 \times 10^{-13}\text{m}. \tag{A.25}$$

Note that the equilibrium emittance depends on the beam nominal energy through γ^2, the bending radii of the dipoles and the lattice functions (betatron and dispersion). The rms energy spread, on the other hand, depends on the nominal energy and bending radii. In machines where no vertical bending occurs, except possibly from quadrupole misalignment or other errors, the lack of vertical dispersion reduces the anti-damping effect of quantum excitation. In these machines, which comprise most electron storage rings, the vertical emittance is much smaller (by a factor of the order of 100 or more) than the horizontal emittance. Machines called 'damping rings' are designed to reduce the emittance of beams injected into *linear colliders*; a small vertical emittance can lead to *luminosities* (see section A.5) that are higher by orders of magnitude [5, 7].

A.3 Light sources [8]

SR sources, now called more generally 'light sources', are based on storage rings and linear structures that can provide much more intense UV and x-rays than conventional lamps and tubes; their dimensions range from a few meters in diameter for sources with 100–300 MeV electron beams to machines with km dimensions and GeV beams.

The figure of merit in many applications of SR is the *spectral brightness* or *brightness* for short (also called *brilliance* in Europe). Brightness is defined as the number of photons emitted by the SR source per unit time, per unit solid angle, per unit area at the source, and per unit bandwidth around a given frequency:

$$\text{Brightness} = \text{Photons}/(\text{s}\,\text{mm}^2\,\text{mrad}^2\,\text{BW}) \tag{A.26}$$

Another important property of SR radiation is the degree of *coherence*. The electrons in a beam bunch emit SR from bending by a simple magnet or from the oscillating trajectories in an insertion device (wiggler or undulator [2]). The SR emitted by individual electrons in single magnets and wigglers adds *incoherently*: if the bunch contains N particles, the total intensity (per unit frequency or *spectral bandwidth*) is simply N times the intensity from an individual particle. In undulators, by contrast, if the electron trajectories at a given pole are all in phase, the resulting intensity of SR will be proportional to N^2; this type of radiation is called *coherent synchrotron radiation* or CSR. Although SR from wigglers is incoherent, they yield more intense SR than single-bend magnets because they are longer and involve multiple bends. Further, since the number of particles in a typical bunch can be very large (10^{10} for a 1.6 nC bunch, for example), the intensity of a coherent source can be many orders of magnitude larger than the intensity of an incoherent one. In reality, the SR intensity always has incoherent and coherent components; their ratio is intimately related to the structure of the beam bunch, i.e. its *line density profile*, and the geometry and dimensions of the storage ring and/or the insertion device. In free-electron lasers (FELs), additional effects occur from the interaction of the emitted SR and the electrons themselves.

The *first generation* light sources were built with high-energy physics research in mind but the parallel use of many of these machines as SR sources developed over time. *Second generation* light sources, by contrast, were dedicated SR sources from the start; their use for multi-disciplinary research often involving multiple beam lines as shown in figure A.1. An example of a second generation light source is the 2 GeV SRS machine at Daresbury, UK. *Third generation* light sources are storage rings optimized for high brightness (see equation (A.26)) and high beam quality, i.e., low emittance, through the use of insertion devices like undulators. Examples are the 1.5 GeV Advanced Light Source (ALS) in Berkeley, CA, and the 3.0 GeV Diamond in Daresbury, UK. The smallest machines of this type produce SR with photon energies below 2 keV (soft x-rays), while the largest ones are designed for hard x-rays.

Finally, *fourth generation* light sources can be defined as those that exceed third-generation light sources by one order of magnitude or more in a parameter like brightness, coherence or (shorter) pulse duration. Fourth generation light sources include storage rings with ultra-low emittance and photoinjector, linac-based FELs that produce VUV to hard x-rays in ultra-short bursts (picosecond and femto-second). Examples of these advanced SR sources are the FLASH FEL facility at

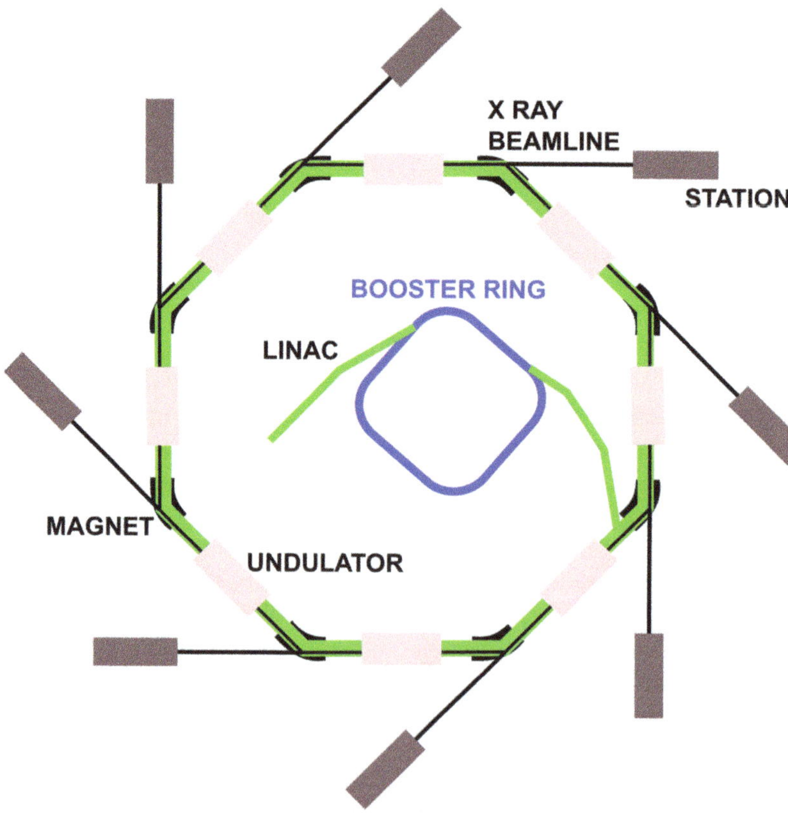

Figure A.1. Schematics of x-ray beamline stations around synchrotron storage ring.

DESY in Germany, the European x-ray FEL facility (XFEL) in Hamburg, the MAX-IV facility in Lund, Sweden, and the Linac Coherent Light Source (LCLS) at SLAC in the US. LCLS utilizes one-third (1-km) of the SLAC linac to accelerate electrons to 14.3 GeV; the 112-m long undulator produces coherent x-rays at photon energies between 0.8 and 8.0 keV corresponding to wavelengths between 1.5 and 0.15 nm. The average brightness is 4.5×10^{22} photons s^{-1} $mrad^{-2}$ mm^{-2}/0.1% bandwidth.

A.4 Basic collider physics

The *center-of-mass* (CoM) energy is more relevant than the laboratory energy in colliders like the 27 km-circumference Large Hadron Collider (LHC). Because of the particle–wave duality of quantum physics, on the other hand, the energy can be related to a corresponding wavelength range: from 0.04 nm (1 keV electrons) to 10^{-4} fm (10 TeV protons—roughly the maximum energy at LHC). The first number corresponds to a fraction of the size of an atom, while the second is much smaller than the proton radius. Thus, electron microscopes, which operate with keV electrons, can peer at much finer details than optical microscopes.

If a particle of mass m (particle 1) and total energy $E_1 = \gamma mc^2$ strikes another particle with the same mass m but *at rest* ($E_2 = mc^2$), then the center-of-mass energy E_{cm} of the two-particle system is

$$E_{cm} = [2(\gamma + 1)]^{1/2}\mathrm{mc}^2. \tag{A.27}$$

(To obtain this result, the *total* laboratory energy squared, $E_1^2 + 2E_1E_2 + E_2^2 - p_1c^2$, is equated to the CoM energy E_{cm}, and equation A.3 is employed.) In non-relativistic collisions of this type, only half the kinetic energy of the incident particle is available, since the CoM moves at half the speed of the incident particle, compared to $E_{cm} \cong \sqrt{2\gamma}\,E_0$, $\gamma \gg 1$, in highly relativistic collisions. By contrast, if the particles collide head on, we have

$$E_{cm} = 2(E_1E_2)^{1/2}, \tag{A.28}$$

or $2E$ if the particles have the same energy. As an example, the 14 TeV CoM energy for colliding protons in the LHC corresponds to a *laboratory-frame* kinetic energy of 10^5 TeV for a proton colliding with another one at rest.

The basic physics scaling for colliders: the *circumference* is proportional to Energy/Magnetic Field = E_{cm}/B, where E_{cm} is the CoM energy and B the magnetic field of the *main bending dipoles*. The length of the circumference (including tunnel and associated infrastructure) is only second to the superconducting magnets in determining the final cost of a collider. Table A.1 summarizes basic data for four hadron circular colliders, two real ones (Tevatron and LHC), one never constructed (SSC), and one planned (FCC-hh).

High-energy colliders require 'pre-acceleration' with other smaller machines. An example is the LHC complex, illustrated schematically and highly simplified in figure A.2. The yellow dots indicate the four largest experiments [9].

Table A.1. Main parameters of four hadron circular colliders.

Accelerator	Momentum (GeV/c)	Circumfer./Avg. radius (km)	Magnetic field (T)	Number of dipoles
Tevatron (Fermilab)	1000.94	6.28/1.0	4.4	744
Large Hadron Collider—LHC (CERN)	7000.94	27.4/4.4	8.33	1232
Superconducting Super Collider —SSC (US)	20 000.94	87.1/13.9	6.8	4168
Future Circular Collider—FCC-hh (CERN)	50 000.94	100/15.9	16	4668

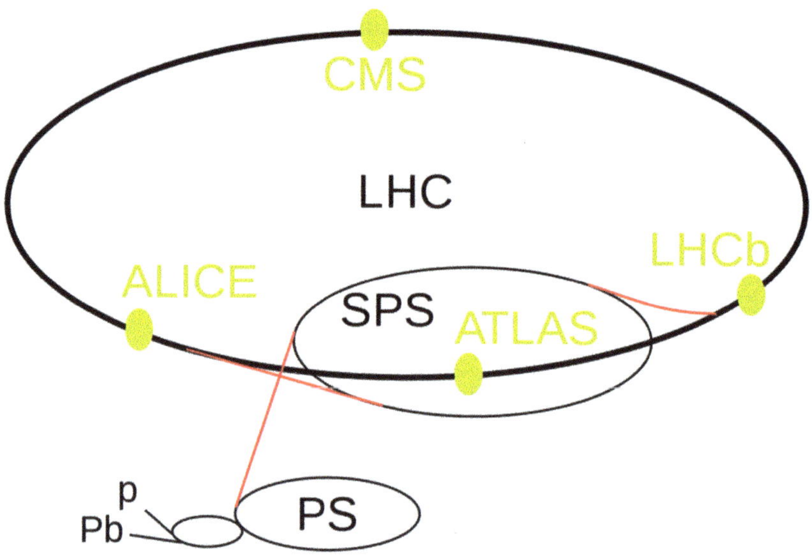

Figure A.2. Schematics of CERN Large Hadron Collider complex. Protons (p) and ions (P_b) start in linear accelerators, followed by a booster ring (small circle), then the Proton Synchrotron (PS), the Super Proton Synchrotron (SPS), and finally the 27-km-circumference LHC [9]. This LHC.svg image has been obtained by the authors from the Wikimedia website where it was made available under a CC BY-SA 2.5 licence https://creativecommons.org/licenses/by-sa/2.5/deed.en. It is included within this book on that basis. It is attributed to Arpad Horvath.

A key parameter in collider physics, other than the CoM energy, is the *luminosity*. It is defined as the event rate, for a specified process, per unit cross-section for that process. Therefore, the actual event rate is equal to the product luminosity × cross-section. Cross-section is measured in ‘barns’, with 1 barn = 10^{24} cm^2. As an example [10], the total (including inelastic collisions and other processes) cross-section for proton–proton collisions at 7 TeV (14 TeV CoM energy) in LHC is 110 mbarn, The luminosity is of the order of 10^{34} cm^{-2} s^{-1}, or about 100 times larger than achieved at the Fermilab Tevatron.

A.5 Computer resources

A free program called ***Radiation2D***, written by Tsumoru Shintake of RIKEN in Japan [11], provides a great educational tool for illustrating the radiation pattern of a moving charged particle following different geometries. For example, the reader can study: (1) the change in electric field lines for different speeds (as fractions of the light speed) of a particle moving uniformly on a straight path; the radiation patterns (again for different speeds) of a particle moving in (2) a circle; (3) a bending magnet; (4) an undulator.

We include in the author's website three MATHCAD worksheets: ***DipoleRadiation.xmcd*** is used to calculate the basic pattern of dipole radiation; ***SynchRad.xmcd*** contains calculations using formulae from section A.2 for electrons, protons and muons (equations (A.10)–(A.12)), and ***AcceleratorCalc.xmcd***.applies basic formulae from collider physics to LHC.

There is also the Android app TAPAs (see appendix B) which is useful for quick calculations of e.g., SR parameters for electron storage rings and insertion devices and others.

Certain sections of text in this chapter have been reproduced with permission from [8]. Copyright John Wiley & Sons.

References

[1] Wollnik H 1987 *Optics of Charged Particles* (New York: Academic)

[2] Bernal S 2022 *A Practical Introduction to Beam Physics and Particle Accelerators* 3rd edn (Bristol: IOP Publishing)

[3] Henderson S, Holmes J and Zhang Y 2009 *Lecture 9a, Synchrotron Radiation* (Nashville, TN: USPAS)

[4] Emery L 2002 Damping rings *Lecture Notes Physics 575—Accelerator Physics and Technologies for Linear Colliders, Winter 2002* (Argonne National Laboratory)

[5] Wolski A 2014 *Beam Dynamics in High Energy Particle Accelerators* (London: Imperial College Press)

[6] Lee S Y 2004 *Accelerator Physics* 2nd edn (Singapore: World Scientific)

[7] Winick H (ed) 1994 *Synchrotron Radiation Sources—A Primer* (Singapore: World Scientific)

[8] Joseph C L and Bernal S 2016 *Modern Devices, The Simple Physics of Sophisticated Technology* (New York: Wiley) http://bcs.wiley.com/he-bcs/Books?action=index&bcsId=10233&itemId=0470900431

[9] https://en.wikipedia.org/wiki/Large_Hadron_Collider

[10] https://lhc-closer.es/taking_a_closer_look_at_lhc/0.luminosity

[11] https://groups.oist.jp/qwmu/software

IOP Publishing

Small Particle Ring Accelerators and Paul Traps

Case studies and prospects

Santiago Bernal

Appendix B

Computer codes

We begin this appendix in section B.1 with a mention of general tools useful for accelerator-related calculations: MATHCAD, SMATH SOLVER (freeware), MATLAB and Python. The sections on 'Computer Resources' at the end of all chapters and appendix A in this book rely mostly on MATHCAD files that can be found in the author's website; equivalent files for the other programs will be available as they are developed and uploaded.

In sections B.2–B.7 we present brief descriptions of several computer codes for designing particle accelerators: MAD, ELEGANT, WINAGILE, TRACE, TAPAS, and a few others. We include hyperlinks for downloading the codes, instructions for installation, and explicit application examples in most cases. Unfortunately, some of the most user-friendly codes such as WINAGILE are now considered 'legacy' and retain value mostly as educational resources.

B.1 General tools: MATHCAD, SMATH SOLVER, MATLAB, Python

MATHCAD and its freeware counterpart SMATH SOLVER (formerly SMATH STUDIO) are self-documenting codes for general calculations in math, science and engineering. Although not as popular as MATLAB or Python, MATHCAD and SMATH SOLVER are convenient tools for simple calculations and others that involve solving algebraic and differential equations, doing numerical or symbolic matrix algebra, special functions, image and signal processing, and programing.

Almost all numerical examples quoted in the 'Computer Resources' sections of the book were written in MATHCAD 14 and 15, with all files available in the author's website, but some SMATH SOLVER files are included as well. In addition, we are uploading MATHCAD PRIME 8.0 or 9.0 versions of the files as they are developed (PRIME 8.0–9.0 has a convenient conversion utility). In figure B.1 we show a MATHCAD 15 document for the solution of the damped harmonic oscillator. For comparison, the equivalent MATLAB and Python codes are presented below.

DAMPED SIMPLE HARMONIC OSCILLATOR

5-29-2015

$m := 1$	Mass in kg
$k := 100$	Spring constant in N/m
$b := 1$	Damping constant in Ns/m
$IC := \begin{pmatrix} 1 \\ 0 \end{pmatrix}$	Initial conditions, x(0) =1 m, x'(0)=0 m/s
$D(t,x) := \begin{pmatrix} x_1 \\ \frac{-k}{m} \cdot x_0 - \frac{b}{m} \cdot x_1 \end{pmatrix}$	Differential equation vector. First element is dx/dt, and second element is d2x/dt2.
$M := 1000$	Integration points
$Sol := rkfixed(IC, 0, 5, M, D)$	Solution array using 4th order Runge-Kutta method with fixed step. First column (0) is time, second column (1) is "x", and third column (2) is dx/dt.
$i := 0..M$	

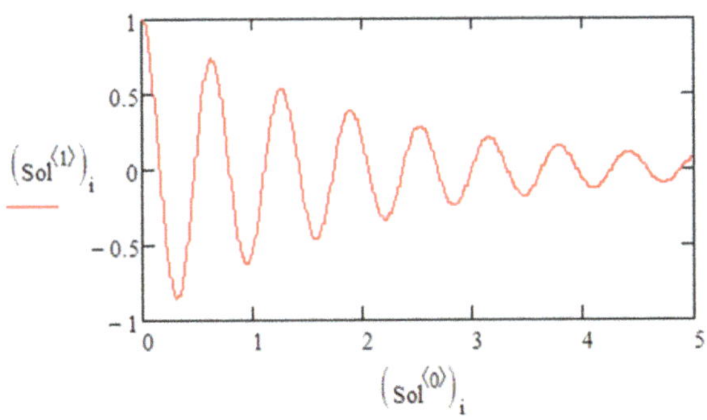

Figure B.1. Simple example of MATHCAD document.

In MATLAB do:

1) Write *and* save the "*M-file*" named `oscillator.m` (use the built-in script editor):

```
function xpp = oscillator(t,x)
m = 1; k = 100; b = 1; % mass, spring and damping constants (SI units)
xpp = zeros(2,1);
xpp(1) = x(2);
xpp(2) = -(k/m)*x(1)-(b/m)*x(2);
end
```

2) In *command* window run

```
[t,x]=ode45(@oscillator, [0 5], [1 0]); plot(t,x(:,1))
```

to generate *x* vs. *t*.

For Python 3.* write (and save as "file.py"):

```
"""Damped Simple Harmonic Oscillator S. Bernal, 5-29-2015"""
from scipy.integrate import odeint
from pylab import *  # for plotting commands
def deriv(x,t): # return derivatives of the array x
   m = 1    # Mass in kg
   k = 100  # spring constant in N/m
   b = 1    # damping constant in Ns/m
   return array([ x[1],-(k/m)*x[0]-(b/m)*x[1] ])
time = linspace(0.0,5.0,1000)
xinit = array([1.0,0.0]) # initial values
x = odeint(deriv,xinit,time)

figure()
plot(time,x[:,0],'r-') # x[:,0] is the first column of x
xlabel('t')
ylabel('x')
show()
```

To run, type "python3 file.py" in an UBUNTU 20.04 terminal.

B.2 MAD-8 and MAD-X

The 'Methodical Accelerator Design (MAD)' program has been used at CERN for over 20 years to simulate, design and optimize large linear and circular particle accelerators. A few major examples are the Proton Synchrotron (PS), the Super Proton Synchrotron (SPS), the Large Electron-Positron collider (LEP), and the Large Hadron Collider (LHC). The old version MAD-8 was 'retired' in 2002, but at least two versions are still available for download: 8.23.06 and 8.51/15 from SLAC and CERN.

MAD does not have a GUI, as it is more a *language* than a code. The User's manual [1] contains useful physics discussions and detailed explanations of all commands, syntax, etc. There is also 'A MAD-X Primer' manual with useful examples [2]. On the other hand, there are a few very useful tutorial presentations available online. We have adapted an example from one of them [3] to use it with the October, 2017 MAD-X release, version 5.03.07. We emphasize that we are only scratching the surface of a very powerful and complex program for single-particle simulations of accelerators. Space-charge effects can be simulated also, but in conjunction with another code called PTC. The CERN website has more details.

DOWNLOAD: To download MAD-X or MAD-8 use http://mad.web.cern.ch/mad/. For MAD-X, click 'Releases' on the left list and then 'releases repository.'

INSTALLATION: for Windows 10, move the madx-win64.exe (64-bit Win) or madx-win32.exe (32-bit Win) file to a folder 'MAD-X'. Rename the file to simply 'madx.exe.'

RUN/EXAMPLE: the MAD-X language is not case sensitive. The file below, FODOUMERrev.MADX, is self-explanatory for the most part. A FODO unit cell for the UMER machine (chapter 1) is defined, and the beta functions and phase advances are plotted for arbitrary initial values `BETX=0.9, BETY=0.7, ALFX=0, ALFY=0`. The calculations are repeated for *periodic* matched conditions ('cell matching' in the terminology of the MAD User's Reference Manual), and for assumed phase advances of 90°. Comments are preceded by //, /*, or !. Notice that the energy 'PC' is in units GeV/c (very small for the 10 keV UMER!).

```
//FILE: FODOUMERrev.MADX
//MADX Example 1: FODO cell
//Author: L. Deniau (CERN), S. Bernal (U. of MD)
//Date: July 6th, 2020

BEAM, PARTICLE=ELECTRON, PC=10158E-06;

D : DRIFT, L=0.10836;
QF: QUADRUPOLE, L=0.05164/2, K1= 168.36;
QD: QUADRUPOLE, L=0.05164, K1=-168.36;

FODO: LINE=(QF,D,QD,D,QF);
USE, PERIOD=FODO;

! Non periodic solution
TWISS, BETX=0.9, BETY=0.7, ALFX=0, ALFY=0;
MYPLT: PLOT, HAXIS=S, HMAX=0.35,
 VAXIS1={BETX,BETY}, VAXIS2={MUX,MUY},
 COLOUR=100, INTERPOLATE;

! Periodic solution
TWISS;
EXEC, MYPLT;

! Match phase advance to 90 degrees
MATCH, SEQUENCE=FODO;
VARY, NAME=QF->K1;
```

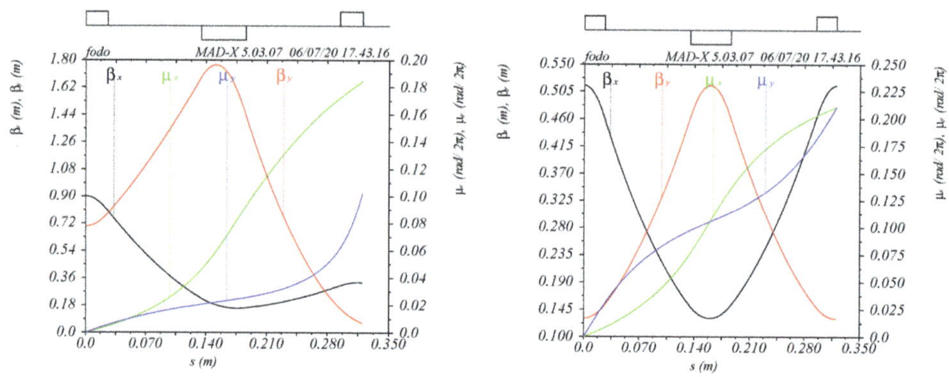

Figure B.2. Output plots from MAD-X: unmatched FODO cell (left), and matched FODO cell (right). Notice the diagrams for the FODO cell on top.

```
VARY, NAME=QD->K1;
CONSTRAINT, RANGE=#E, MUX=0.25;
CONSTRAINT, RANGE=#E, MUY=0.25;
LMDIF, TOLERANCE=1e-20;
ENDMATCH;

! Matched solution
TWISS;
EXEC, MYPLT;
```

The program is run by typing

```
madx FODOUMERrev.MADX
```

The output is in the form of the postscript file 'madx.ps' which can be opened with GhostView in Windows. In the example above, three plots (one per page) are generated. The first two plots are shown in figure B.2. Additional examples can be found in [4].

B.3 Elegant

ELEGANT is a powerful third-order matrix code widely used to model light sources and other accelerators, from the Advanced Photon Source (APS at Argonne National Lab) to X-ray FELs such as the Linac Coherent Light Source, LCLS, at Stanford National Lab. ELEGANT was started by Michael Borland from APS in 1988 and is maintained by him and others at APS and elsewhere [5, 6]. There is also the 'ELEGANT forum' where users can usually find solutions to code or model-related problems. The code is freely available from the APS website for all major platforms, but you will need to register with your e-mail to get a key to download the files and documents (ELEGANT, SDDS, examples, and manuals).

DOWNLOAD: You can download all the files from the APS website: https://www.aps.anl.gov/Accelerator-Operations-Physics/Software.

INSTALLATION: For the Windows installation, you will need to install CYGWIN and MS Visual C++. Follow the steps in the APS website. Installation in Ubuntu Linux is relatively straightforward; we recommend building the 'rpm' packages using the script in Step 1 instead of trying the pre-built RPMs. For MacOS, download the binary executables for both SDDS and ELEGANT (darwin-x86 compressed folders). ELEGANT will run with at least MacOS 12.3 Monterey in machines with the new M1 chip or the Intel chip.

RUN/EXAMPLE: the main files in ELEGANT are the *input* file with extension '.ele' and the *lattice* file with extension '.lte'. These files can be edited with any text editor such as WordPad in Windows, 'gedit' in Linux, or TextEdit in the Mac. Below is an example of a *lattice file*, 'umer1.lte', corresponding to UMER (chapter 1).

```
!umer1.lte
!Simple Example from UMER

DR1: DRIF,L=0.05418
QD1: QUAD,L=0.05164,K1=-153.419
DR2: DRIF,L=0.03538
DIP1:                    SBEND,L=0.0376,ANGLE=0.1745329252,
E1=0.0872664626,E2=0.0872664626,HGAP=0.0287,FINT=0.3,&
FSE=-0.0
DR3: DRIF,L=0.03538
QF1: QUAD,L=0.05164,K1=153.419
DR4: DRIF,L=0.05418
DR5: DRIF,L=0.05418
QD2: QUAD,L=0.05164,K1=-153.419
DR6: DRIF,L=0.03538
DIP2:                    SBEND,L=0.0376,ANGLE=0.1745329252,
E1=0.0872664626,E2=0.0872664626,HGAP=0.0287,FINT=0.3,&
FSE=-0.0
DR7: DRIF,L=0.03538
QF2: QUAD,L=0.05164,K1=153.419
DR8: DRIF,L=0.05418

!Define Beamlines

FODO1: LINE=(DR1,QD1,DR2,DIP1,DR3,QF1,DR4)
FODO2: LINE=(DR5,QD2,DR6,DIP2,DR7,QF2,DR8)
SECTION: LINE=(FODO1,FODO2)
RING: LINE=(18*SECTION)
```

The notation and structure of this file are like those of files used in the MAD codes (previous section). UMER consists of 18 sections, each having two FODO periods. There are two bend dipoles and four quadrupoles per section (see chapter 1 for additional details.) 'K1' represents the quadrupole focusing constant κ_0 in m^{-2}; the bend dipoles are rectangular 10° bend magnets (two 5° edge angles E1 and E2), and FINT is a constant ('field integral') that is introduced to implement edge effects (see TRANSPORT manual [7] for definition.) Note that '!' must precede comments. The ELEGANT manual [6] describes in detail a great number of accelerator lattice elements and their parameters.

The ELEGANT *input file* is 'umer1.ele':

```
!umer1.ele
 !Simple Example from UMER

 &divide_elements
 maximum_length=0.002
 name = *
 &end

 &run_setup
 lattice = umer1.lte
 default_order = 3
 rootname = umer1
 use_beamline = "ring"
 output = %s.out
 sigma = %s.sig
 p_central = 0.199
 final = %s.fin
 magnets = %s.mag
 &end

 &run_control
 n_steps = 1
 n_passes=5
 &end

 &twiss_output
 matched = 1,
 beta_x = 3.119e-1,
 alpha_x = 1.324,
 beta_y = 3.090e-1,
 alpha_y = -1.315,
 eta_x = 0.0,
```

```
etap_x = 0.0,
eta_y = 0,
etap_y = 0,
filename = %s.twi
&end

!&bunched_beam
bunch = %s.bun
n_particles_per_bunch = 10000,
emit_x = 8.048e-6,
emit_y = 8.048e-6,
beta_x = 3.119e-1,
alpha_x = 1.324,
beta_y = 3.090e-1,
alpha_y = -1.315,
eta_x = 0.0,
etap_x = 0.0,
eta_y = 0,
etap_y = 0,
use_twiss_command_values = 0,
Po = 0.199,
sigma_dp = 0.01,
sigma_s = 1.69,
distribution_cutoff[0] = 2.5
distribution_cutoff[1] = 2.5
distribution_cutoff[2] = 1.73
distribution_type[0] = "gaussian"
distribution_type[1] = "gaussian"
distribution_type[2] = "hard-edge"
save_initial_coordinates =1;
!&end

!&track &end

&save_lattice filename = %s.new
output_seq = 1
&end

&stop &end
```

Notice that the 'umer1.ele' file above is divided into blocks that begin with '&name of block' and end in '&end'. For example, we have '&run_setup', '&twiss_output', '&run_control', '&bunched_beam', '&track', and others. The elements of all available blocks are explained in detail in the ELEGANT manual.

Further, note how the lattice file and the beam line are called in the '&run_setup' block. In addition, an entire block can be commented out by using '!', as is the case with the '&bunched_beam' block above.

The code is run by typing

elegant `umer1.ele`

In the example above, the code finds the matched lattice functions Betax, Betay, Etax and Etay, i.e., the transverse betatron and dispersion functions. To plot Betax, for example, type

```
sddsplot umer1.twi -col=s,betax
```

Notice the command 'sddsplot'; it is part of the SDDS tools, an important component of ELEGANT. To print the betatron tunes one would type

```
sddsprintout umer1.twi -par={nux,nuy}
```

In order to track and plot *beam* functions, as opposed to *lattice* functions, we need the '&bunched_beam' block in the '. ele' file. Thus, we specify the initial beam parameters, distribution types in *x, y, z* and number of particles (10 000 in the example.) As an example, we type

```
sddsplot umer1.sig -col=s,Sy
```

to plot the rms envelope in the vertical plane as a function of the reference-orbit coordinate '*s*'. Figure B.3 shows the output plots of Betax with `use_beamline = "ring"`, and `use_beamline = "section"` in the '&run_setup' block.

Running ELEGANT and plotting lattice and beam functions and doing complex operations can be made very convenient by writing SDDS scripts. For example, we can have a simple script, 'simple_script.txt' that will run 'umer1.ele', and display the

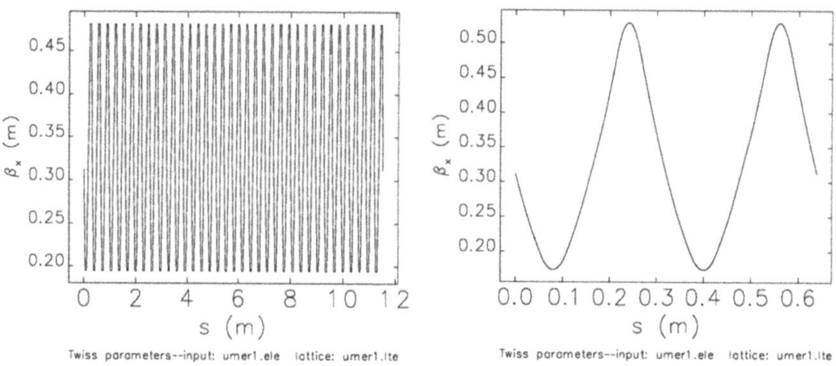

Figure B.3. Horizontal betatron functions calculated in ELEGANT: one turn or 18 sections (left), and one section (right) of the University of Maryland Electron Ring (UMER).

betatron tunes, horizontal and vertical chromaticities, momentum compaction, and plot Betax, Betay:

```
sddsmakedataset settings.sdds \

elegant umer1.ele

sddsprintout umer1.twi -par={nux,nuy}
sddsprintout umer1.twi -par={dnux/dp,dnuy/dp}
sddsprintout umer1.twi -par=alphac
sddsplot umer1.twi -col=s,betax
sddsplot umer1.twi -col=s,betay
```

The script can be run by typing './simple_script' (no quotation marks) in MS Windows (under Cygwin), but the text file must be first converted to a Unix text file using the command 'd2u simple_script.txt' ; 'd2u' stands for 'dos-to-unix.' Such conversion is not required in Linux or MacOS. Finally, a very useful command in ELEGANT is 'sddsquery filename'; for example, 'sddsquery umer1.twi', 'sddsquery umer1.sig' will list all parameters that are part of the lattice and beam sets, respectively.

B.4 WINAGILE

WINAGILE is a free, (old) MS Windows-based, computer code for beam transport calculations in transfer lines and rings. It can be used for particle tracking, beam envelopes, closed-orbit and corrections, resonance studies, and many other applications [8]. Both incoherent and coherent space-charge effects can be included as well, but the models are not very accurate. The code was written by P J Bryant from CERN, who is also the author, with Kjell Johnsen, of *The Principles of Circular Accelerators and Storage Rings* [9]. With some guidance, running the basic WINAGILE can be learned in a matter of minutes. The GUI is not the best in the world, and more than a few bugs remain in the available version (ver. 3.5), but it is nonetheless a working code which can be easily applied to model many accelerators.

DOWNLOAD: You can download the code from the author's website at https://www.terpconnect.umd.edu/~sabern/PIBPPA-3rdEd/Appendix%20A The file 'package(ver3.5_D054).zip' includes the windows-executable file, lattice examples (extension 'LAT') and help file 'winagile.hlp.'

INSTALLATION/RUN: the code will run under MS Windows XP, or under WINE *(PlayOnLinux)* in Ubuntu Linux. It is possible to run WINAGILE in Windows 10 but only inside a virtual Windows XP machine through VMware Player or Oracle's Virtual Box.

EXAMPLE: thinl1l2.lat—this lattice has two thin quadrupoles of opposite strengths and separated by a distance shorter than their individual focal lengths $f = 0.1$ m, in magnitude. The total length of the beam line is 1.0 m.

(1) Run WINAGILE and open the example file **thinl1l2.lat** (this you will find in the author's website.) This lattice contains two types of elements: **DRIFT** and **THINQ.**
(2) Note that you are in the **MAIN WINDOW.** Explore the different column values. Only the 'Length' and 'Thin quad kL' columns have values. In particular, the length of THINQ is zero, by definition. The integrated strengths (in m^{-1}) of the quadrupoles have opposite signs.
(3) In the 'Calculations' menu choose 'Transfer Line.' A window 'Twiss Input' opens; click OK. The input here is irrelevant to our present calculations. You end up with a new window called **BEAM PARAMETERS**.
(4) In the 'Calculations' menu choose 'Tracking'. Enter 0.005 for 'Horizontal position, x' and also for 'Vertical position, z'. The window now changes to **TRACKING**.
(5) In the 'Graphs' menu choose 'Trajectories'. A window with a blue background opens showing the horizontal and vertical trajectories. The default particles are 1 GeV protons, but this is irrelevant, as all the information for the optics is contained in the 'kL' values. (**MAIN WINDOW**.) Click 'Save' and then 'Back' to return to the 'Tracking' window.
(6) Repeat (4) with −0.005 values for x and z. Select the entire field to be able to change the signs from the original values (this is but the first example of a bug in the GUI!—but be patient.)
(7) Choose 'Trajectories' again in the 'Graphs' menu and then 'All.' You will see the wonders of 'alternating gradient' focusing: net focusing on both planes! You can now try increasing the separation of the quads to be more than their individual focal lengths. To do this you need to go back to the **MAIN WINDOW**, then 'File', 'Edit'. Double-click on the field that you want to change and press OK. Then go to 'Check Data' and select either option. You're ready to repeat the tracking calculations.

We had presented in chapter 1 (see figure 1.10) a calculation of closed orbits and corrections in UMER using WINAGILE. Furthermore, the lattice examples that come with the code include the second generation light source (see appendix A, and [10]) National Synchrotron Light Source Vacuum Ultraviolet, NSLS VUV, at Brookhaven Natl. Lab in Upton, NY, the Low-Energy Anti-Proton ring, LEAR, at CERN, the Proton Storage Ring, PSR, at Los Alamos Natl. Lab. and a few other machines and demonstration lattices.

B.5 TRACE 2D, 3D

The best description of the TRACE programs comes from the Abstract to the TRACE 3D documentation [11]: 'TRACE 3D is an interactive beam-dynamics program that calculates the envelopes of a bunched beam, including linear space-charge forces, through a user-defined transport system. TRACE 3D provides an immediate graphics display of the beam envelopes and phase-space ellipses and accommodates fourteen different types of fitting or beam matching options.' The documentation includes several appendices with excellent concise discussion of basic topics such as the σ-matrix, Twiss parameters, mismatch factor and others. Furthermore, a summary of TRACE 2D commands is given in appendix K. TRACE 2D is used for unbunched, i.e., continuous or coasting beams.

DOWNLOAD: the codes are freely available from Los Alamos National Laboratory after a simple registration process. See http://laacg.lanl.gov/laacg/services/download_trace.phtml. The codes will also be available from the author's website.[1]

INSTALLATION: the .zip file is decompressed to reveal the executable file 'trace2d.exe' or 'trace3d.exe' which can be moved to the desktop in Windows. The examples and input files can remain in any folder, but the Windows system folder 'Program Files (x86)' is not recommended.

RUN/EXAMPLE: clicking on the TRACE icon will produce a file-opening page. In our example, the input file is 'fodotest.t2d' (TRACE-2D). A blank page is produced after opening the file, but simply typing 'g' (for 'graph') will render the beam horizontal and vertical phase- space ellipses at the input (left pane) and output (right pane) planes, together with the beam line and the transverse beam envelopes at the bottom. See figure B.4.

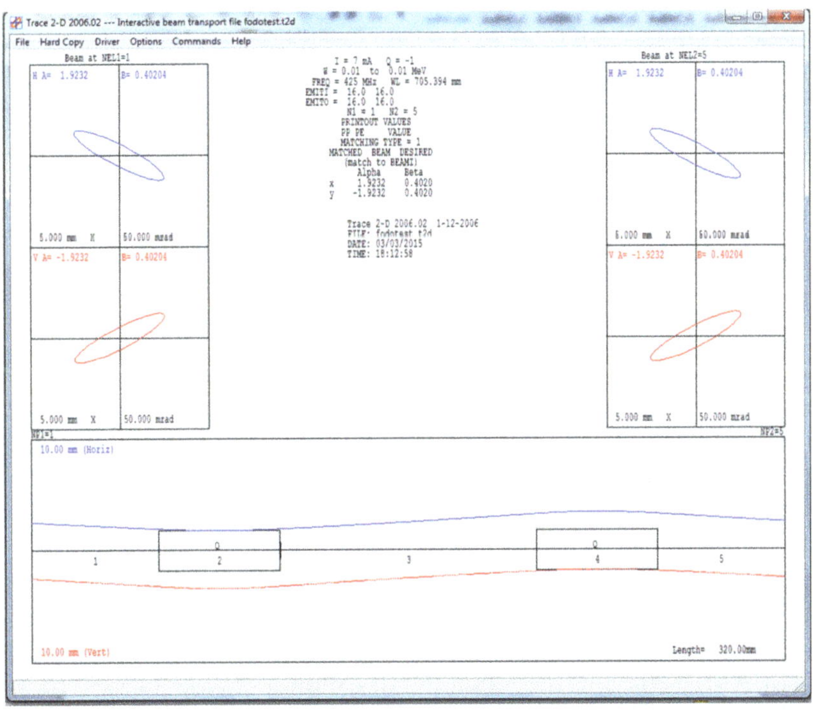

Figure B.4. TRACE 2D window after matching the beam envelopes in one cell of UMER for a 10 keV, 7 mA, 16 μm (x and y edge emittances) electron beam.

[1] https://ireap.umd.edu/clark/faculty/1273/Santiago-Bernal

```
&DATA
ER= 0.511 Q= -1. W= 0.010 XI= 7.000
EMITI = 16 16
BEAMI = -0.914037 1.8 0.914037 1.8
BEAMF = 0.00000 1.8 0.00000 1.8
BEAMCI= 0.00000 0.00000 0.00000 0.00000
FREQ= 425.000 PQEXT= 2.50 ICHROM= 0 IBS= 0 XC= 0.0000
XM= 10.0 XPM= 50.0000 YM= 10.00
XMI= 5.0 XPMI= 50.0000 XMF= 5.0 XPMF= 50.0000
N1= 1 N2= 5 SMAX= 0.5 PQSMAX= 2.0 nel1= 1 nel2= 5 NP1= 1
NP2= 5
MT= 1 NC= 4 MP=0,000 0,000 0,000 0,000
MVC=0,000,0 0,000,0 0,000,0 0,000,0

CMT(001)='DRIFT' NT(001)= 1 A(1,001)= 54.18
CMT(002)='Quad' NT(002)= 3 A(1,002)= 0.57049E-01 51.64
CMT(003)='DRIFT' NT(003)= 1 A(1,003)= 108.36
CMT(004)='Quad' NT(004)= 3 A(1,004)= -0.57049E-01 51.64
CMT(005)='DRIFT' NT(005)= 1 A(1,005)= 54.18

WS(1)= 0.00000 0.00000 0.00000 0.00000 0.00000 0.00000
0.00000
WS(8)= 0.00000 0.00000 0.00000 0.00000 0.00000 0.00000
0.00000
SIGI(1,1)=   45.216   -186.84   0.00000E+00   0.00000E+00
0.00000E+00
SIGI(1,2)=-186.84    851.67    0.00000E+00    0.00000E+00
0.00000E+00
SIGI(1,3)=0.000E+00 0.000E+00 34.734 57.21 0.0000E+00
SIGI(1,4)=0.000E+00 0.000E+00 157.21 815.21 0.0000E+00
SIGI(1,5)=0.000E+00 0.000E+00 0.000E+00 0.000E+00 131.21
COMENT='FODO match for 7.0 mA, 10keV'
&END
```

The second line of 'fodotest.t2d' defines the beam: 'ER' is rest mass in MeV, 'W' the kinetic energy in MeV, and 'XI' the beam current in mA (continuous beam). The next line contains the initial transverse edge emittances (EMIT) in mm-mrad. BEAMI (BEAMF) contains the initial (final) Courant–Snyder (Twiss) parameters α and β in the horizontal and vertical planes. XMI, XPMI, XMF, XPMF define the scales (in mm) for the initial and final phase-space ellipses. XM, YM define the scales for the envelopes. There are five elements in the beam line, three drift spaces and two quadrupoles; the drifts are in mm, and the quadrupole parameters are the gradient in T/m, and hard edge effective length (51.64 mm in our case).

The matching type is MT = 1, for ellipse parameters α_x, β_x, α_y, β_y. Envelope matching is carried out by typing 'm' and then 'g'. The phase advance and Courant–Snyder parameters are displayed by typing 'f'. A list of commands is provided in a dropdown menu. The input file can be modified and run again without saving or exiting the program. As an example, the current can be set to zero by typing 'i', and then 'xi=0' in the new window; notice how the beam is mismatched. After re-matching, the new phase advances can be obtained; comparing these values with those obtained previously with full beam current illustrates the effect of linear transverse space charge (see chapter 1.)

B.6 TAPAS

TAPAS is an app for Android devices. From the Google Play website—'TAPAs stands for Toolkit for Accelerator Physics on Androids. It allows 'back-of-the-envelope' accelerator physics calculations, with more than 20 inter-linked types of calculations. These include storage ring scaling with energy and circumference, longitudinal dynamics in rings, free-electron lasers, electron linear accelerators, electron guns, bunch compression, iron-dominated multipole magnets, short pulse X-rays, undulator effects on orbit and optics, undulator radiation properties, undulator models, top-up parameters, electromagnetism, synchrotron radiation, particle passage through matter, and engineering calculations.' See also [12, 13].

TAPAS is free and can be found in the Google Play Store; search for 'TAPAS Borland'. Alternatively, use the QR code in figure B.1 of [13].

Figures B.5(a) and (b) show screens from TAPAS for the fourth generation 3 GeV light source MAX-IV, which is located at Lund University in Lund, Sweden. Several

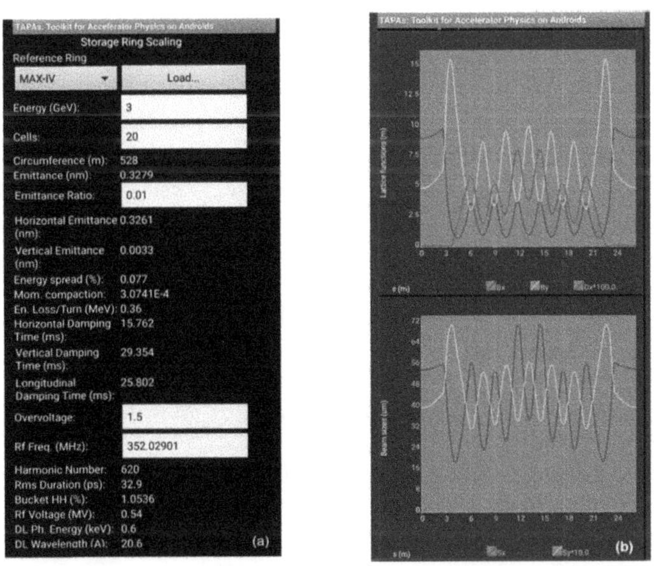

Figure B.5. TAPAS screenshots for the 528-m circumference storage ring MAX-IV: (a) scaling, and (b) lattice functions (top) and beam sizes (bottom). Reproduced courtesy of Michael Borland..

parameters shown in figure B.5(a) were defined in section A.2 on synchrotron radiation (e.g., energy loss per turn, and damping times).

B.7 Other codes

There are two codes used for cyclotron and cyclotron-like accelerators, i.e., FFAG or FFA accelerators (chapter 4), that were used in an FFA school in 2019 at the Paul Scherrer Institute (PSI) in Switzerland: ZGOUBI and OPAL. See https://indico.psi.ch/event/7313/page/1707-ffa-school. ZGOUBI can be downloaded following the link (424 MB). Additional useful information on ZGOUBI, OPAL and similar codes can be found in [14].

References

[1] Grote Hans, Roy Ghislain, Schmidt Frank and Deniau Laurent (ed) 2022 *The MAD-X Program (Methodical Accelerator Design), Version 5.08-01, User's Reference Manual* (Geneva Switzerland: European Laboratory for Particle Physics) http://mad.web.cern.ch/mad/releases/5.08.01/madxuguide.pdf

[2] Herr W and Schmidt F 2004 *A MAD-X Primer, CERN-AB-2004-027-ABP* (Switzerland: Geneva)

[3] Métral E 2013 *Training-Week in Accelerator Physics* (Sweden: Lund)

[4] Murphy J 1993 *Synchrotron Light Source Data Book, Version 3.0*42333 (revised) Informal Report Brookhaven National Lab., BNL

[5] Borland M 2000 *Technical Report No.* LS-287 Argonne National Laboratory

[6] Borland M 2021 User's manual for elegant, program version 2021.4 *Advanced Photon Source*

[7] Carey D C, Brown K L and Rothacker F 1995 Third-order TRANSPORT—a computer program for designing charged particle beam transport systems SLAC-R-95–462, Fermilab-Pub-95/068 414

[8] Bryant P J 2000 AGILE, a tool for interactive lattice design *Proc. of EPAC 2000 (Vienna, Austria)* 1357

[9] Bryant P J and Johnsen J 1993 *The Principles of Circular Accelerators and Storage Rings* (Cambridge: Cambridge University Press)

[10] Bernal S 2022 *A Practical Introduction to Beam Physics and Particle Accelerators* 3rd edn (Bristol: IOP Publishing) Section 8.2

[11] Crandall K R and Rusthoi D P 1997 TRACE 3D documentation *Los Alamos National Laboratory Report*LA-UR-97–886 Los Alamos National Laboratory 3rd edn

[12] Borland M 2013 Android application for accelerator physics and engineering calculations *Proc. of the 2013 Particle Accelerator Conf. (Pasadena, CA)* 1364

[13] Borland M 2016 Recent improvements to TAPAS, the android application for accelerator physics and engineering calculations *Proc. of the 2016 North-America Particle Accelerator Conf., NAPAC2016 (Chicago, IL)* pp 625

[14] Smirnov 2017 Computer codes for beam dynamics analysis of cyclotronlike accelerators *Phys. Rev. Accel. Beams* **20** 124801

www.ingramcontent.com/pod-product-compliance
Ingram Content Group UK Ltd.
Pitfield, Milton Keynes, MK11 3LW, UK
UKHW051848100926
8556IPUK00005B/518

9 780750 348898